拳法

KENPO :

AN ILLUSTRATED INSTRUCTOR'S MANUAL

.

WITH RESTRAINT AND RESUSCITATION TECHNIQUES

井口松之助著
INOGUCHI MATSUNOSUKE

シャハン・エリック訳
ERIC SHAHAN

COPYRIGHT © 2020 ERIC MICHAEL SHAHAN
ALL RIGHTS RESERVED.
ISBN-13 : 978-1-950959-18-1

正三位伯爵鷲尾隆　聚公題字

正三位子爵海江田信義君題字

日本中學校長杉浦重剛先生序題

早繩

活法拳法教範圖解 全

柔術教師久富鐵太郎先生序文

演武舘長今泉八郎先生跋文

安達吟光先生摸寫

井口松之助著述

Kenpo :
An Illustrated Instructor's Manual With Restraint and Resuscitation Techniques

Introductory Calligraphy by

- Count Takashi Washio
- Viscount Nobuyoshi Kaieda
- Junior High School Principal Nagasugi Urashige

Introduction by
Jujutsu Instructor Hisatomi Tetsutaro Sensei

Additional Comments by
Head of Enbukan Dojo Imaizumi Hachiro Sensei

Illustrated by
Ando Gin Sensei

Written by
Inoguchi Matsunosuke

Translator's Introduction

Several years ago I translated and self-published a Jujutsu manual called *The Police Officer's Essential Illustrated Guide : Kenpo*. The book, which was originally published in 1888, was originally only available to police officers however later in that same year it underwent a revision and was made available to the general public. A decade later, Inoguchi Matsunosuke consulted with the original author, Hisatomi Tetsutaro, and reformatted the book two more times, finally releasing an edition with new illustrations and expanded explanations. The same 16 Torite, or Police Arresting Techniques, are presented in both the 1888 version and the 1899 version, however the focus was shifted away from a police only method to general self-defense. The final version included both Kappo, or resuscitation, and Haya-nawa, or rope restraining techniques. The final title of this book was :

An Illustrated Guide to Jujutsu Training Including Fast Tie and Resuscitation : Also Known as Police Kenpo

While this book uses words like Jujutsu, Kenpo and Torite they are all basically interchangeable.

Jujutsu : Soft method

Kenpo : Way of the Fist, also means Jujutsu with striking

Torite : Jujutsu or Kenpo techniques focused on seizing and restraining people.

Note :

This book begins with three sets of calligraphy done by various luminaries. Each calligraphy work is a phrase broken up over 3 ~ 5 pages. An introduction of the artist and an explanation will follow each, however in the original book only the artists' names and titles are given.

牧成

自得

正三位隆衆

鷲尾隆聚 **Washi-no-o Takatsumu 1843?-1912**

技成自得

The Acquisition of Skill is Entirely up to You

Washi-no-O was a soldier and politician active in the late Edo and early Meiji Era. He was was active in the Boshin War and met with Saigo Takamori, urging him to end his campaign against the government. An avid sword practitioner he founded the *Sword and Spear Arts Preservation Society* 剣槍柔術永続社 with Yamoka Tesshu in 1884. This group laid the foundations for the later *Greater Japan Martial Virtue Society* 大日本武徳会 established in 1895. Washi-no-O was married twice and had 10 sons and 4 daughters. He died in 1912 at the age of 71.

海江田 信義 Kaieda Nobuyoshi 1832-1906

Kaieda was a Samurai from the Satsuma Domain. He was a student of 示現流 Jigen School Kenjutsu and also studied 薬丸自顕流 Yakumaru Jigen School which featured the Nodachi, a sword with a blade length over 90 centimeters.

He was famous for being involved in the Namamugi Incident, also known as the Richardson Affair.

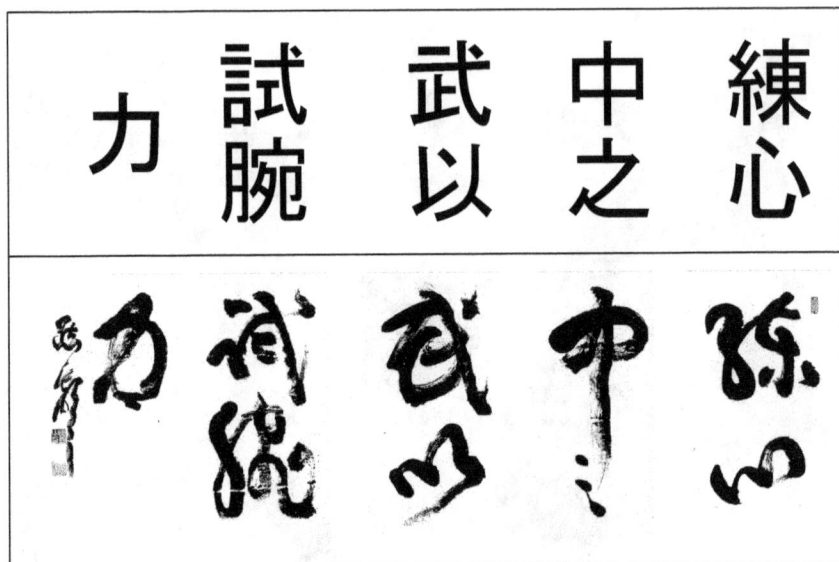

練心中之武以試腕力

As you refine your spirit use martial arts to test the strength of your arms.

Note :
Namamugi Incident/The Richardson Affair.

The incident occurred in 1862 in the vicinity Nagamugi Village near Kawasaki. A British man named Charles Richardson was cut down by Samurai of Kagoshima and several of his companions were wounded. There are several versions of what happened but apparently Richardson's group encountered a group of Samurai and were not sufficiently respectful. The killing sparked outrage and even resulted in the castle town of Kagoshima being bombarded by the British Navy.

Kaieda was said to have either lead the attack on Richardson, struck the struck the "Tome" killing blow, or both. He would have considered the final blow to be a merciful one, to end the man's suffering.

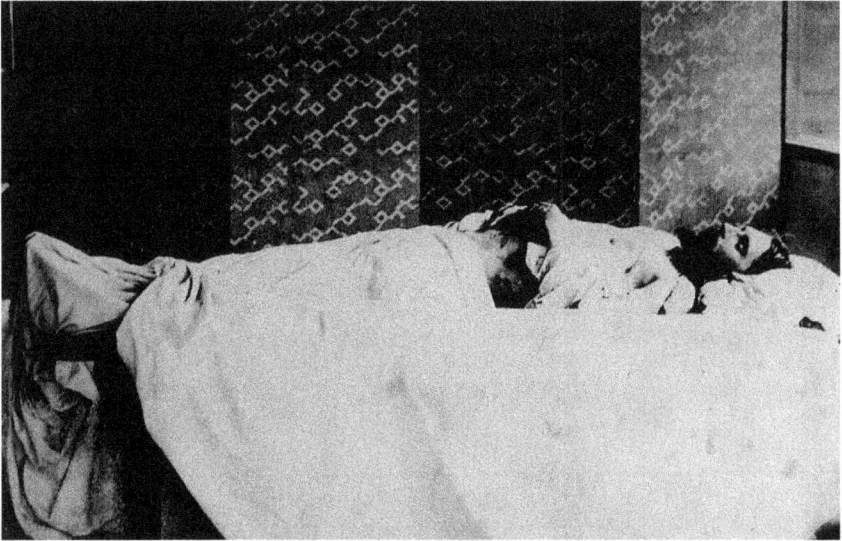

Photo of Richardson's body

The Nagamugi Killing by Haykawa Shozan late 19th century

無恃其不来

恃吾有以待

也

右孫子之語

天公監道士題

Sugiura Shigetake 杉浦重剛 1855-1924

Sugiura was an educator and thinker born in Shiga Prefecture. He was a dedicated student and in 1873, at the age of 18, was selected to demonstrate a science experiment in front of the Meiji Emperor. In 1876, while studying at a university in Tokyo, he was he had the chance to study overseas. He attended University in Great Britain from 1876-1880. He later established the Tokyo English School and managed the private school Shokojuku, focusing on youth education.

Sugiura's calligraphy is a quote from the Art of War :

故用兵之法，**無恃其不來，恃吾有以待之**；無恃其不攻，恃吾有所不可攻也

*The art of war teaches us to **rely not on the likelihood of the Attacker's not coming, but on our own readiness to receive him;** not on the chance of his not attacking, but rather on the fact that we have made our position unassailable*

拳法教範圖解序

活法
早繩

老生前年警視ニ奉職ノ頃大ニ感スル處アリ從

來嗜ム處ノ柔道ヲ以警官ナル人ヲシテ常ニ講

セシメ惡漢不呈之徒ニ會シ職權執行ノ際ニ備

エン事ヲ測リ當時長官ニ具申ス長官之ヲ嘉シ

採用スル處トナリ則各署員ヲシテ公務ノ餘暇

講習セシム爾來其功不少然ルニ職員ノ内從來

其技ニ長スルモノ多ク八其流派ヲ尊奉シ我流

Introduction
By Hisatomi Tetsutaro

Several years ago while serving as a police officer I (Note : Tetsutaro refers to himself as "this old fellow") began developing an idea. A person seeking to become a police officer should be well versed in a subject I have been passionate about for a long time, Judo. If they are equipped with Judo and encounter neer-do-wells, vagabonds or ruffians they can employ these techniques in their capacity as an officer of the law. I presented a detailed plan to the Police Commissioner at the time and it was enthusiastically adopted. Henceforth, when officers were not on duty they would receive training in Judo. The results were not insignificant. Amongst the officers there were many who had extensive Judo training and were very loyal to their respective martial arts school. This dedication to their particular school sometimes prevented training seminars from proceeding smoothly.

ヲ講習スルニ圓滑ナラサル處アリ老生之ヲ憂
ヒ各員之内三拾餘名ヲ本署ニ會シ其法ヲ論シ
各得意ノ技ヲ講セシメ内一二ヲ拔擢シ十六種
ノ形手ヲ編製シ各署員ヲシテ之ヲ講セシメ依
テ大ニ協和スルヲ得タリ故ニ老生之ヲ一ノ書
册トナシ名付テ拳法圖解ト稱シ各位ニ頒ツ爾
來今日ニ至レリ然ルニ井ノ口氏始メ該術ノ師
家今泉先生此技ヲ修正再刊センフヲ老生ニ請

Seeking a solution to this while respecting their art I called together 30 or so of them at the Police Headquarters. We discussed the various methods and each person demonstrated what they considered their best techniques and we selected one or two of these from each school. The result was a program consisting of 16 different Torite : Police Arresting techniques. When the final program was presented to everyone, it was received with great enthusiasm.

In the end I was able was able to compile a book with all these techniques. I called it *The Police Officer's Essential Illustrated Guide : Kenpo* and it was published in 1888. Copies were made available to every police officer. Several years later, I was contacted by a Mr. Inoguchi along with the head teacher of another school, a Mr. Imaizumi Sensei, who had made some corrections to each of the techniques and inquired about the possibility of reissuing the book.

此書タルヤ素ヨリ一時和同ノ爲メ編製セシ
モノニシテ後來ニ流傳ス可キモノニアラス然
リト雖モ其請所切ニシテ止マス依テ後人ノ笑
ヲ顧ス今泉氏ヲ始メ井ノ口氏ト會シ二三ノ修
正ヲナシ其好意ノ在ル處ニ任ス讀者諸君此意
ヲ了セラレンコヲ希望ス

明治三十一年五月

久冨鐵太郎識ス

26

Since the original volume was a peaceful compromise between all the parties involved I was initially reluctant to agree. However, with the changes they proposed, henceforward all the necessary teachings that need to be transmitted are now included for those that follow. Thus I abandoned my reluctance, a decision that will no doubt bring smiles to the faces of future generations.

I made two or three corrections, but I left the rest up to the good intentions of Mr. Imaizumi and Mr. Inoguchi. It is my sincere wish that the readers of this volume understand the teachings contained in this volume.

May Meiji 31 (1898)
Hisatomi Tetsutaro

Notes :

Hisatomi Tetsutaro 久富鉄太郎 (?-1899?)

The 7th generation head of Shibukawa School Jujutsu, a branch of Sekiguchi school.

Imaizumi Hachiro 今泉八郎 (?-1900)

Imaizumi was a practitioner of Sekiguchi Shin-Shin School Jujutsu and Kusunoki School Kenpo. Originally he was a student of Imaizumi Kumataro, however after extensive training he was adopted by the master and continued as the next Imaizumi. He later left Edo to train in Tenshin Shinyo School, and was able to attain the highest level. Following this he learned Araki School Bojutsu. Eventually he combined all the teachings he learned and opened his own Shin Kage School Jujutsu 神蔭流柔術. In the 16th year of Meiji he became the Jujutsu Shihan (head instructor) for the Shitaya Section of Tokyo Police Department. It was said he had over 5,000 students.

早縄活法 拳法教範圖解

編者 井口松之助

緒言

該書ノ起稿ハ明治拾七年ヨリ同拾九年凡三ヶ年間警視廳ニ於テ其頃久

富鐵太郎先生ヲ初メ各柔術家拾六流議ヨリ丗有餘名ノ世話掛ノ妙手ヲ

撰擧シ其後既ニ久富先生書冊ニ成サシタレ共充分ニ至ズ故ニ今回更ニ

同先生ノ望ミノ儘ニ綴リ我全國一般ノ警官軍人ニ及諸學校教師生徒ニ

至モ此形ヲ廣大ニ弘メ度キ一フ余ニ教傳ノ話ヲ受タル故余ハ嚮肆ヲ

以テ營業チナスト云共幼少ヨリ武術ヲ好ミ幸ニ而テ久富先生ヲ初メ今泉

八郎先生其他各先生ニ教導ヲ受テ尚蘭師安達吟光先生腕ヲ振ヒ蓮テ摸

寫ナナシ筆ニ徙難共處ハ盡以テ解ショク其見易キ爲前後左右ヨリ寫生

ナシ數日ヲ以テ綴リタル拳法形ト早縄活法ニ至モ記載ナシタリ最

モ余ハ先ニ柔術劍棒圖解ニ及武道圖解秘訣又柔術極意教授圖解ヲ出版

シ最近ノ書ニハ 死掘自在 技竹鞭法 柔術生理書ヲ著シ大高節ヲ博シ尚 神刀流 剣舞前篇 劍舞圖解

Editor's Introduction

Police Officer Mr. Hisatomi first began compiling this book in the 17th year of the Meiji Emperor, 1884. For the next three years Hisatomi Tetsutaro gathered Jujutsu practitioners from 16 different schools. The 30 or so renowned martial artists he selected agreed on a set of techniques which Mr. Hisatomi compiled into a book.

Since there were some elements lacking in the original volume we have reformatted it while maintaining Hisatomi Sensei's original vision. This book is now being made available not only to police and military officers but also teachers and students all over Japan. This expanded distribution will allow for a greater number of people to receive instruction in Jujutsu. Copies will also be available in bookshops. Fortunately, thanks to this book, children will now have the chance to develop an affinity for Budo. All this is due to the book written by Hisatomi Sensei and revised by Imaizumi Sensei and all the other instructors of classical martial arts. In addition, through the drawing done by master illustrator Ando Gin (Note : rhymes with "kin" not the alcohol) complicated actions are clearly depicted throughout.

Ando Sensei was able to complete his sketches over the course of just a few days. The result is easy to understand drawings. This included not only Kenpo techniques but also Haya Nawa "fast tie" restraining techniques as well as Kappo resuscitation techniques.

The first book Inoguchi Sensei wrote was called *An Illustrated Guide to Jujutsu, Kenjutsu and Bojutsu* as well as the follow-up volume *Illustrated Secrets of Martial Arts*. Later he published *Tenshin Shinyo : Illustrated Instructor's Guide to the Inner Mysteries of Jujutsu*. Recently he has just published *Jujutsu, the Living Art : Freely Using Resuscitation and Jujutsu Massage*, which has received high praise. Inoguchi Sensei is completing the soon to be released volume called *Kenbu (Sword Dance) An illustrated Instructor's Guide to Solo Training*.

致範ヲ出版シタレバ余ガ著ス出版書ハ他ノ編輯者ト違ヒ實地活用的ナ
旨トシ諸學校生徒ニモ一讀スレバ圖解ヲ見ルバカリニテモ先生不入ノ
獨學ニ出來様組立アル故ニ此擧法ハ軍人警官ニ及町道場ノ諸先生ニ至
ル迄モ開讀スレバ忘レタル處ハ覺ヘ又武術ヲ知サル者ハ千萬ノ力ヲ得
ル護身術ノ助ニナルフヲ僅ノ時日ヲ資テ卒業スルノ法ヲ以テ余ガ苦心
ナシタル處ヲ講讀諸君ヨ猶余ノ心切實地ヲ障セラレンフヲ望ム　最モ
此書ヲ以テ解シ難キ者ハ軍人警官ニ限リ無謝儀ニテ敎導ナスフヲ許
ス　下谷區同朋町角今泉演武舘神田區錦町二丁目吉田柳眞舘ニ於テモ
著者ガ敎授ナスフ得最近出版物ハ各流秘傳劍術極意敎授圖解ヲ出版ナス尚
此書中ニモ警視ノ劍術ノ形アリ出版ノ切ハ尚高評ヲ乞　余ハ最モ無學
ニシテ文章ノ熟セザル處ニ及字句ノ誤リ重言等モ有之ハ實地的ヲ專務
ニナス故ニ諸君之ヲ諒セヨ

明治三十一年四月下旬

源　義　爲　謹　白

Unlike many of the writers of today, Inoguchi Sensei has many years' experience training in these arts and this is reflected in his writing. This book is arranged in such a way that if students were to read it just once and study the images, they could easily learn without a teacher. Thus *Kenpo : An Illustrated Instructor's Manual* is not only for soldiers and police officers but also for the Sensei teaching at local Dojo. By reading this book you will recall lessons long forgotten. Those with no knowledge of martial arts will find the power of a thousand divisions of 10,000 soldiers added to your self-defense.

My dear readers this is a method that teaches you to achieve competency in a short amount of time. If you wish to further develop yourself by receiving friendly instruction or you find some part difficult to understand, your questions will be happily answered. This offer of course applies not only to members of the armed forces and police, but the general public as well. The authors teach at :

- Imaizumi Bukan Dojo
  Shitaya Ward Tomo Machi Kado

- Yoshida Yanagi Shinkan Dojo
  Kanda Ward Nishiki Town, Ni-Cho Me, Tokyo

In addition, I would like to mention that the soon to be released *Secret Teachings of All Sword Schools : Illustrated Instructor's Guide to Kenjutsu* will contain official police Kenjutsu, sword fighting, techniques. To be sure this volume will receive no small amount of praise once it is released.

Finally, though you can no doubt perceive as much from my writing I am not particularly adept at brushing introductions. Though no doubt I have made numerous errors in this passage, I would like to ensure the reader that I have done extensive practical training and I beg you to overlook my mistakes.

Minomoto Yoshitame

體勢圖解 正面直立之圖

總テ柔術ノ致ハ體勢ヲ旨トシテ 此ノ體勢トハ常ニ稽古ヲナス時ニモ
體ノ崩亂ザル樣ニ體育法ヲナスベシ 最モ實地ニテ出合トモ稽古ニ其心
ナクバ不意ニ掛ラル時ハ其効ナシ 故ニ圖解ノ如ニナスベシ

體勢法ニ曰直立

ヂナシ口ヲ結ヒ

ヲ胸ヲ開キ兩手

ヲ歪シテ拇指

掌ニ折圓テ腰ヲ

張リ 下腹ニ力ヲ

入充分ニ氣合込

テ攝ヘタル處ヲ眞之位ノ第一圖トモ云ベキ處ナリ又常ニ歩行スル時

モ稽古ヲナス時モ左リ足ヨリ徐々ト進出ルベシ 又傘法ノ形ハ最モ鑿

32

Illustration of Taisei : Body Positioning
Shin no Kurai : Front View of "True Standing Position"

When teaching Jujutsu, the first thing teachers focus on is Taisei, or body positioning. Whenever you are doing Keiko, training, you should remember the rules about positioning your body so that you do not collapse or have erratic movement. If you encounter a situation in real life and have not taken to heart lessons about body positioning then will be unprepared. Thus I have included an illustration along with an explanation.

According to the Taisei-ho you should be standing straight with your mouth tied (closed) and your chest out. Your arms should hang at your sides with your thumb tucked against your palm and other four fingers wrapped around it.

Flex your hips and focus your power in your lower abdomen. Your body should be filled with fighting spirit. This is Shin no Kurai and is shown in the illustration. Whether you are walking down the road or advancing on an opponent during Keiko, start with your left foot and move it forward slowly and carefully.

視ノ形故ニ（サーアベル）ヲ帯テ稽古スベキ所ナレ共町道場等ニテハ木

太刀ヲ用ル拳法ニテ太刀ノ必用ノ處ハ柄取・柄止。柄搦。又見合取ニテ

ハ凶器等ニ小太刀ヲ用ユル　形ノ切ニハ受方即チ甲者又我レト奮ス處

アリ取方チ乙者又敵ト記ス處モアリ總テ柔術ハ速氣々合勇氣ノ三氣ヲ

以テ發聲ナナス者ナリ甲ヨリ陽ノ聲ニテ（エイヤ）ト發音ノ聲

出時ニハ乙ニ於テ又陰ノ聲ヲフクミテ（ナー）ト答ナナスベシ　甲ノ

陽ノ發スル聲トハ形ノ切ニ敵ト對顔ナナシ直立ノ體敵ヲ白眼テ下腹ニ

充分之勇氣ヲ滿チ（エイヤト）口ヲ開キテ發スル掛聲ヲ陽ト云又乙ハ同對

顔シテ下腹エ力ヲ入白眼テ又掛聲ヲ甲ヨリ發セハ（ナー）ト口ヲ結ヒテ

答ヘルヲ陰ノ聲ト云ナリ最モ掛聲ハ氣合ノ増ス者故ニ此ノ處ニ記載ス

ルナリ流名ニ依テハ無聲ノ處モアリ成共初心ノ爲ニハ掛聲ヲ肝要トナ

ス最モ總テ武術ニハ（エィヤ）（ヤー）（トー）（エィー）等劍術柔術居合棒長巻

其外ニ又投ルニモ打込ニモ發聲ヲ掛クベキ者ナリ故ニ之ヲ記シヌ。

Since these Kenpo Jujutsu techniques are part of the official police curriculum, you should train with a saber in your belt.

Note : The Japanese police carried western style one-handed sabers at this time.

At the same time local Dojo often train Kenpo using a Kidachi, wooden swords. The reason a Tachi, sword, is an essential weapon for training is so that the following techniques can be trained :
Tsuka Tori, Tsuka Tome, Tsuka Karame as well as Miai Tori.

For these techniques a Kodachi is used. When techniques are being described the Uke-gata, the receiver of the technique, also referred to as Kosha, former, or simply written as "you." The Tori-kata, the one performing the technique, will be referred to as Otsusha, latter, or simply Teki, the "Attacker." There are three kinds of Ki or spirit in Jujutsu :

Soku k i : Fast Spirit
Kia i : Vocal Challenge
Yuk i : Concentration of Martial Spirit

The first Kiai, *Eiya!*, is done by you and is known as Yo no Sei which means Yin Voice (Sei could also be read as Koe.) The Yin Kiai shouted by you indicates that you are beginning the technique. The Attacker responds with In no Sei, Yang Voice, of *O-h!* Your posture should remain straight as you face the Attacker and glare into his eyes with Hyaku-gan, a white-cold stare. Your Yu-Ki, the essence of your bravery, should fill the area below your belly. Next open your mouth and shout a Kake-goe of *Ei-Ya-Toh!* This is a Yo no Sei, Yin Voice. In response the Attacker looks you in the eye with Hyaku-gan and focuses his power under his belly. He shouts a Kake-goe of *O-h!* with his mouth firmly shut. This is referred to as Yang Voice.

The purpose of Kake-goe is to increase your Kiai or fighting spirit. Thus I have written about it here. Depending on the school there are Musei no Tokoro, or places where you don't use Kake-goe, however it is essential that beginners use Kake-goe. All martial arts use some variant of *Eiya! Yaa! Toh! Ei-i!* They appear in Kenjutsu, Jujutsu, Iai, Bojutsu, Nagamaki Jutsu and so on. While there are Kake-goe used when throwing and striking they will not be introduced in this volume.

體勢圖解　其二體　搆圖其一

體勢其二ノ圖解ニ曰總テ柔術家ニ於テハ最モ要務トスベキ者ハ身搆ヲ
專一ニスルベシ其法ハ圖ノ
如ニ兩手ヲテ我陰嚢ヲ圍ヒ
中腰ノ樣ニ兩足ヲ横一文字
ニ開キ瓜先ヲ外部ヘ向ケ腰
サ据ヘ口ヲ結ヒ下腹ニカヲ
入寫生ノ搆ヲ眞ノ位其二ト
モ云ベキ處ナリ最モ此腰ニ
テ腰投。　入腰脊頁投等形亂
捕ニハ最自由自在ニ配動ナ
ナスベキ搆ヘナレバ常々腰
ナタメスベシ

36

Taisei Zukai
Illustration of Body Positioning #2
Karada Gamae Zu Sono Ichi
Body Kamae Illustration #1

This is Taisei Sono Ni, illustration of Body Positioning, Stance #2. Amongst Jujutsu practitioners this is considered to be the most fundamentally important stance. You should pay close attention to this explanation.

As the illustration shows, both hands should surround Kintama, the groin, and your hips should be in Naka-goshi, hips held low, with your legs in Yoko Ichi Monji. Yoko Ichi Monji describes a straight line extending out from your left and right. Your toes should be facing outward stabilizing your hips. Your mouth is "tied shut" meaning your lips are pressed together as you place all your power in your lower abdomen. This illustration shows what is known as Shin no Kurai #2, True Stance #2.

This stance is also known as Koshi Nage, Hip Throw. This stance gives you the greatest freedom of movement when moving in on an opponent, doing a Sei-o Nage, Oi-Nage or doing Randori "free-sparring." You should make moving from this stance part of your everyday training.

此ノ圖ノ構ハ前圖構ヘヨリ左リ足ヲ一尺二三寸後ヘ引テ膝ヲ突キ右足ヲ立膝ナナス腰ヨリ上ハ前ト同足ノ趾ヲ我カ肛門ヘ押當ルベシ

我レ投ラレテ起上ルニモ此樣ニテ起テ眞ノ位ノ其三共云フパシ最モ敵ト對顔シテ白眼アイチナシ發聲ノ音ノ出ルモ起上リノ勇振殘心附ルモ此圖ノ樣ニテ形ノ都合ニ依テ足ハ左右アリ此ハ右立膝ヲ記スナリ最モ發心、中心、殘心氣合雄氣ノ滿ツル此ノ構ニアル處ナレバ圖ノ解釋ヲ詳細ニ言ヘバ膝ト爪先ヲ外部ヘ向ケテ横一文字ニ開キ膝頭ト足先ノ並ブ樣ニナシ下腹ニ力ヲ入ルベシ

Taisei Zukai
Sono San
Karada Gamae Zu Sono Ni

#3
Illustration of Body Positioning
Illustration of Kamae #2

This illustration shows a position that continues from the previous illustration. From that stance, drop your left foot back 1 Shaku and 2 or 3 Sun, 36 ~ 39 centimeters, and plant your left knee on the ground. Your right leg remains upright. From the waist up, everything is the same as in the previous Kamae. The heel of your left foot should be pushing into your anus.

When recovering after being thrown you should end up in this Kamae. This stance is also known as Shin no Kurai no Sono San, True Stance #3. You should be facing the Attacker and glaring at him with "white eyes." When in this stance not only should you do a Kake-goe just as if you were standing but also remember to fill yourself with martial power and keep Zanshin, a state of being fully aware of your surroundings.

Note : The word Zanzhin is comprised of the Kanji 残 Remaining and 心 Heart. Every school has its own definition of Zanshin but generally it is to remain aware and "in the fight" and conscious of the space around you even after a technique is over. Some schools, like the Itto School, define Zanshin as, "have no doubts in your movement."

While this illustration shows your left knee on the ground, depending on the situation you could have either your right or left knee on the ground. Thus your resolution, focus, Zanshin, spirit and bravery should all inhabit this Kamae. They should fill your being. Regarding details of this Kamae I would like to note that the toes and knee of the leg that is upright should be facing outward like in Yoko Ichimonji Kamae. Your toes should be furthest out with your knee next. Put power in your lower abdomen.

距離 圖解

對顏白眼ノ圖

距離ト八其柔術道塲ニ及

稽古ナス塲所ノ廣狹ヲ量

双方共形ノ切ハ此禮ナナ

シ一時元ノ處ニ開キ甲乙

ノ發聲ト同時ニ中央迄進

ミ出凡三尺斗間ノ明圖ノ據

ナナスチ距離ト云同時ニ

左足ヲ壹尺二寸斗左ヘ斜

ニ双方共ニ開キ直ニ乙ヨ

リ仕掛ルベシ最モ拳法内

ニハ甲ヨリ先ニ仕掛ル處

Kyori, or distance, is determined by the amount of space you have in a Jujutsu Dojo or training space. Before beginning a technique both training partners move toward each other do a Rei, bow of respect, and then return to the starting point and stand in Kamae.

Note : "The starting point..." seems to be the furthest distance the Dojo space will allow.

Then, at the same time, Koh-Otsu (Former and Latter,) or you and the Attacker, do a Kake-goe (shout) and approach the center of the training area. When you and the Attacker have approached within 3 Shaku, 90 centimeters, of each other, take one of the Kamae (stance) shown in the illustration. This is what is meant by Kyori, or Distance. Then, at the same time both of you step out 1 Shaku and 2 Sun, 36 centimeters, to the left with the left foot. This opens you up diagonally to each other.

The Attacker should immediately begin an attack. Generally, in Kenpo if the Koh-sha (you) attack first it is with Eri-nage (Collar Throw,) Yo no Hanare (Moving Away from Yin,) or other such techniques. Another example is when you are not facing the Attacker. This happens in Left and Right Yuki Zure, Ato-tori and other such techniques. The rest of the techniques begin from the Kamae shown in the illustration.

八

ハ襟投、陽ノ離、對顔セザル處ハ左右行連、後捕等ナリ　餘ハ總テ此ノ

圖ノ構ヘニテ割出ス總テ柔術ノ形ニ於テハ同形ノ内ニ左右アル所モ有

成共大略ハ人ノ利キ方ハ右故ニ右ノ形多クアリ　此ノ警視ノ形ハ僅ニ

拾六本ナル故人五度學ベバ我十度モ稽古チナシ一ヶ月間迄ニ卒業ノ出

來ル者ナレハ充分ニ氣合ヲ入勇氣ヲ込テ身體ヲ輕ク我レヲ投ル、共其身

手ヲ打早ク返ルベシ又畫ニ記シアル點線ハ其變化ナル故此替ヲ充分ニ

舊畫ヲ見合シテ其上ニテ稽古スベシ最モ指ノ動キ眼玉ノ附處ニハ畫師

ノ動キナリ故早縄活法ニモ其心ヲ附ベシ最モ活法ノ動キハ充分ノ氣合

ガナクテハ其功ヲ要セス　此形ハ同ジ仕掛ノ處モアリ其形ニ付一本毎

ニハ袴模様ニテ甲乙ヲ見ルベシ見出シニ左右前後斜メ等ノ記載ヲ見合

テ頁讀アレ當形ハ久富先生ノ修正ヲ以テ出版ナシタレバ其見難キ處ハ

著者ノ考案ニテ諸先生ノ試驗ヲ受テ出版ノ梓ニノセタレバ實地ニ手ヲ

以テ敎ヘル樣ニ圖畫ト筆ニ・マカシテ記載アル者ナレバ講讀諸君ノ幸ヒ

ヲ祈ル所ナリ

There is a left and right version to all Jujutsu techniques. For the most part people are right handed, which is why most techniques begin on the right. There are only 16 techniques in the Police Department curriculum. Thus if a person trains the techniques 5 times over 10 Keiko (training) sessions it is possible to complete the course and graduate in a month. This will mean having put sufficient Kiai (martial spirit) Yuki (spirit of bravery) into your body. Your body should be light and even if thrown or struck be able to rapidly respond.

In addition the dotted lines in the artworks indicate changes in movement, therefore you should pay special attention to the details in both the text and the illustrations when training. The illustrator has taken pains to show the way the fingers move and the place the eyes are fixed. This applies to the sections on Haya Nawa (Fast Rope Restraint) and Kappo (Resuscitation) as well. It is particularly important with regards to Resuscitation since without sufficient Kiai (martial spirit/dedication to task) the techniques will fail to revive the subject.

There will be some techniques that seem similar, however you will be able to discern between Koh-Otsu (Former and Latter,) or you and the Attacker, in each technique by the different design of their Hakama. The text has also been adjusted to reflect the art, so the text may be on the left, right, top or bottom to make for easier reading.

These techniques were all corrected by Hisatomi Sensei before publication. For sections we, the writers (Inoguchi & Imaizumi,) felt were unclear we consulted with Sensei from various schools and they tried out and approved the techniques before we published this edition. We took the learning from these sessions in the Dojo and used brush and illustration to introduce them to our readers. It is the dearest wish that readers take this illustrated book and learn from it on their own wherever they can find space to train.

柄取 其一

右ヨリ寫圖

甲方ハ木太刀ヲ帶
ナ進ミ出ル（安ミ乙ニ
此方ト共ニ出シテ我シ乙）
ノ容ト共ニ）双方氣合充
分滿チテ對顏ス同
時ニ左足ヲ一尺二
寸斗リ左斜メニ開
キ（双方其）我ハ左手
（同シ）
拇指ニテ劍ノ鍔ヲ
押ヘテ下腰ニ力ヲ
入テ圖ノ如クニ構
ヘルベシ〻直ニ次
圖ヘウツルベシ

44

1
Tsuka Tori : Taking the Handle of Your Sword
Sono Ichi : Step One
The Attacker is on the Right

You approach the Attacker with a wooden sword in your belt. You should shout a Kake-goe at the Attacker, the response by the Attacker will increase both combatants' martial spirit. Both combatants are filled with fighting spirit as they face off against each other. At the same time you and the Attacker both step diagonally left 1 Shaku 2 Sun, 36 centimeters, opening your body up to each other. At this point you place the thumb of your left hand on the Tsuba, sword guard, of your sword.

Pushing down you place power in your hips as you lower your center of gravity. You should be positioned as shown in the illustration. Step Two is on the following page.

Note : The author uses the word Ki-Dachi, wooden Tachi, or longsword. All information in smaller font within brackets is included in the translation.

其二　取　柄ヲ

敵ハ双手ニ右ニ甲方ノ柄ニ右ニ奪圖ノ如
足ヲ踏出シ拔カント柄ヲ攫ミ引キ同時ニ右足ヲ

甲方ハ敵ノ動勢ト同時ニ力ヲ込ヲ敵ノ眼中ヲ見ハ劍ノ鍔ヲ搨ヘ次圖ノ指ニ捫ヲ額ニ押シ正シ下腹ニ

Tsuka Tori : **Taking the Handle of Your Sword**
Sono Ni : **Step Two**
The Attacker is on the Right

The Attacker steps forward with his right foot one step and, at the same time, grabs hold of the handle of your sword with both hands. He then tries to pull it out. You realize what the Attacker is attempting to do the moment he begins his attack. You hook your thumb over the top of the sword guard to stop your sword from being taken. You put power in your lower abdomen and stare directly into the face of the Attacker. The technique continues on the following page.

柄取 其三

左ヨリ寫圖

甲方ハ充分ノ氣合ヲ込テ敵

ハ劍ヲ引秘カント左リ足ナ

一足引テ抜掛ルヲ直チニ右

手刀ヲ以テ敵ノ眼中ニ霞ナ

入同時ニ右足ナ一足ノ引

ニツレ右足ヲ蹈出ス 敵ハ

此ノ手刀ヲ除ルヲ爲ニ左ヘ顔

ヲ向ケル我レハ直チニ點線

ノ如ク柄頭ヲ下ヨリ握リ腰

ニ力ヲ入テ充分ニ氣合ヲ込

テ次圖ヘ續ク（手刀ト「五指ヲ揃ヘテ

打「ナ云感ハ形ナリ）

（最モ柄取ノ肝要ノ處ナリ）

Tsuka Tori : **Taking the Handle of Your Sword**
Sono San : **Step Three**
The Attacker is on the Left

You should be completely focused with martial spirit, as the Attacker is trying to pull your sword out. He has stepped back with his left foot while pulling. You respond immediately by stepping forward with your right foot and thrusting your right hand out in a Shuto, Sword Hand, aimed at the Attacker's eyes. The Shuto is formed by bringing all 5 fingers together and striking. This distracting attack causes Kasumi, an obscuring mist, to cover his eyes.

The Attacker will flinch away from your Shuto by turning his head to the left. You use this chance to grab the Tsuka-gashira, Pommel of your sword, with your right hand from below. This is shown with the dotted lines. As you do this, lower your hips and build up your martial spirit for the next move, shown on the next page. This set-up is the most important part of Tsuka Tori.

取柄

引下ラント欲スルヲ下ニ引
右膝頭ヘ引
柄ヲ我レ膝頭ヘ
敵ノ手ヲ持ッテ我レ柄ヲ
直ニ敵ハ柄ヲ引附テ又
共ニ引附テ敵ガ右柄頭ヲ
頭ニ引附テ共ニ足ト
後ヘ足ト共ニ足ニ故ニ
圖ヨリ右ニ引ケル故
二圖ニ充分
ニ引

リ
最モ敵ノ背部ハ
手ハ敵ノ兩手ヲ
手ヲ擲ムベシ
處ニ我圖ノ如クヲ擲ムベシ

50

Tsuka Tori : **Taking the Handle of Your Sword**
Sono Yon : **Step Four**
The Attacker is in Front

This continues from step three. Immediately pull the pommel down and over towards your right kneecap. At the same time you should step to the side with your right foot sufficiently. The Attacker continues to hold onto the handle of your sword, trying to pull it free so when you step with your right, he follows one step forward.

Immediately grab the back of his left hand with your right hand. The technique continues on the next page. Ideally you would be able to hold both of the Attacker's hands.

柄取 其五 左斜メ四リ返図

敵ヲ充分ニ引附テ直
ニ下腹ヘ力ヲ入敵ノ
體ノ崩レタル處故ニ
右足ヲ左足ノ處ヘ寄
セナガラニ圖ノ如ク
左手ハ鍔ニシカト持
兩足蹈揃ヘ直ニ右手
ハ敵ノ左手ト柄頭ニ
附テ敵ノ腰ヨリハルト
同時ニ敵ハ右手ヲ放
ス故ニ直ニ次圖ノ如
クニナスベシ　最モ敵ノ左手ハ逆ニ成シ故ニ充分ニ投ラルヽナリ

Tsuka Tori : **Taking the Handle of Your Sword**
Sono Go : **Step Five**
The Attacker is in front on the Left

Having pulled the Attacker towards you sufficiently, place power in your lower abdomen. The Attacker has lost his balance at this point so you bring your right foot up beside your left. Having done that you should be positioned as shown in the illustration. Your left hand should be holding the Tsuba firmly as you bring your feet together. Your right hand should be holding the Attacker's left hand as well as the pommel of your sword.

As the Attacker loses his center of gravity he will let go with his right hand. As soon as he does this, do as is shown on the following page. Since you have reversed the Attacker's left wrist, you will be able to throw him quite a distance.

ハ力ヲ入レズ氣合ト藝トニテ出

來ルコヲ忘ルベカラズ　起上ル

同時ニ甲ト對顏スルヲ殘心ト云

是ヲ柄取ノ形　終ル

〇此形ハ　元天神眞揚流ト眞蔭

流ヲ修正ナセシ形ナリ　久富鐵

太郎先生ト今泉八郎先生驗查ニ

テ今泉榮作先生ト著者井口松之

助ノ形捕圖

54

柄取 其六

正面ヨリ寫図

我レ左リ足ヲ後ヘ大キ
ク引開ク同時ニ右ヨリ
敵ノ手モロトモニ下ヘ
返シ敵ヲ投放スナリ
圖ノ如クノ搆ヘニナル
ベシ　敵ノ起上リ殘心
附ル迄テ此ノ搆ヘヲ崩
スベカラス？敵モ手ガ
逆ニ成ル故ニ投ラルヽ
成バ身ヲ輕クシテ手ヲ
打テ倒ルベシ　總テ形

Tsuka Tori : **Taking the Handle of Your Sword**
Sono Roku : **Step Six**
The Attacker is on the Left, being thrown towards the viewer

Take a big step back with your left foot opening your stance wide. At the same time, pull the Attacker's left hand from right to left and down. This will throw the Attacker, causing him to release. You should end up in the Kamae shown in the illustration, taking care to maintain Zanshin, careful awareness even after the technique is finished, until the Attacker stands. The Attacker was thrown due to pressure on his reversed left wrist. He should not tense up when thrown, but remain loose. The Attacker slaps the ground to indicate he is defeated.

You should always remember to focus on not muscling your way through techniques but on maintaining martial spirit and technical application. After recovering from being thrown, the Attacker should face you and lock eyes showing his Zanshin.
This is the end of Tsuka Tori : Taking the Handle of Your Sword.

About this technique :

This technique has been adapted from Tenshin Shinyo School and Shin Kage School. Hisatomi Tetsutaro Sensei, Imaizumi Hachiro Sensei, Imaizumi Eisaku Sensei and author Inoguchi Matsunosuke all contributed to this arresting technique.

柄止 其一

双方共進出ル形ハ前條ノ一圖
ノ如ク　敵ヨリ甲方ノ柄頭ト
右手首ヲ掴ミ圖ノ如キ樣ニナ
リ受方ハ同ク前ノ一圖ノ如ク
攣ヘテ下腹ニ力ヲ入テ敵ノ眼
中ヲ見ル　乙モ力ヲ入テ甲方
ノ顏ヲ見ナガラ次圖ヘ續ク
ベシ

58

2
Tsuka Tome : Stopping the Handle of Your Sword
Sono Ichi : Step One
The Attacker is on the Left, in front of you

Both combatants approach each other as shown in the previous technique. The Attacker grabs the Tsuka-Gashira, pommel, of your sword with his right hand and your right wrist with his left hand. This is shown in the illustration. You should respond as described in the first illustration in the previous technique, by putting power in your lower abdomen, hooking the thumb of your left hand over your Tsuba, or sword guard, and looking the Attacker straight in the eye. The Attacker is staring you in the eyes as he readies himself. The technique continues on the next page.

Note : The author uses three different terms for "you" : Koh, Uke-kata and Ware (you.)

柄　止

右斜メニ窩圖

敵ハ直ニ左斜メニ左リ足ヲ
蹈出スト同時ニ双手ニテ甲
方ノ右ノ方ヘ押附ル故ニ我
レモ又右足ヲ右ヘ一尺餘斜
メニ開キ　下腹ニ力ヲ入テ
止ルベシ　此時ニ敵ノ體少
々崩レル故ニ直チニ次圖ニ
ウツルベシ

（○此ノ柄止次圖ノ處ガ肝要ノ處
ナリ）

Tsuka Tome : **Stopping the Handle of Your Sword**
Sono Ni : **Step Two**
The Attacker is on the right, slightly diagonally ahead of you

The Attacker steps diagonally to the left and, at the same time, pushes both your hands to your right. You respond by stepping about 1 Shaku, 30 cm, diagonally forward and to the right. Put power in your lower abdomen and stop yourself from being pushed. Since the Attacker will be slightly off balance by this motion, immediately do as shown in the following illustration. This next illustration is the key move in this technique.

柄　止　其三　見所ノ正面ヲ寫圖

甲方ハ直ニ右足ヲ敵ノ前ヘ踏出ス同時ニ躰ヲ進〆敵ノ右手ニテ握リタ
ル處ノ柄ヲ我レヨリ鍔ニ拇指ヲ掛タル儘
下ヨリ左ヘ迴上ルト右ニ持タル敵ノ手解
ル故ニ左手モ同時ニ敵ノ下ヨリ高ク上レ
バ敵ノ指解ケル故直ニ敵ノ左ノ手首ヲ握
リ敵ノ左腕ヲ我レヨリ巻込第四ノ圖ノ如
ノ手ニテ充分ニ持バ敵ノ體崩レル故ニ右
足ヲ引テ次圖ニ續ク

柄　止　其四　正面ヨリ寫圖

敵ノ手ヲ卷込ミ同時ニ右足ヲ一足引テ直
チニ其處ヘ膝ヲ突左足ヲ横ニ開キテ卷込
タル手首ニ左ノ手ヲ添敵ノ手ヲ圖ムベシ

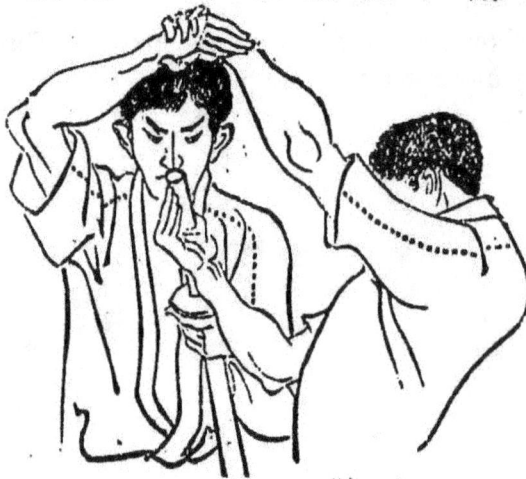

62

Tsuka Tome : **Stopping the Handle of Your Sword**
Sono San : **Step Three**
The Attacker is in Front of You

You should immediately step in front of the Attacker with your right foot. At the same time move your body forward. The Attacker's right hand is gripping the handle of your sword. Keeping your thumb on the Tsuba, drop the sword handle down and then rotate it to the left and up. This will cause the Attacker to lose his grip with his right hand. At the same time raise your right hand up high, which will cause the Attacker's left hand to lose its grip.

Immediately grab the Attacker's left wrist and wrap up his left arm. If done properly then the Attacker will lose his balance and you can drop your right foot back to take him down as shown in illustration 4 on the next page.

敵ハ臆ト痛ノ／ハ崩ト頁チ／亂ニテ双方／ニシ心チリ／終ル殘チハ／共ル此形ノ／○附形ノ勢内／盈川流キ先形／手瀬ノ内リ／車取合富ナ／立合久當リ／チナ修正試／生最モ正驗／スモ又スナ／形ニ最試リ／ナナスモ／リリ圖験

***Tsuka Tome* : Stopping the Handle of Your Sword**
***Sono Yo*n : Step Four**
The Illustration Shows the Scene From the Front

As you wrap his arm up pull your right foot back one step, then immediately plant your right knee on the ground. Your left leg should extend out to the left. Bring your left hand to join your right hand, covering the Attacker's left hand.

This technique can be painful since the Attacker is completely discombobulated. Having been defeated, the technique ends and however, both you and the opponent should maintain Zanshin until the end.

About this technique :

This technique is a Shibukawa School technique called Te-Tuzuki Seisha Tori no Uchi Tachiai, which has been adapted by Hisatomi Sensei. These illustrations reflect those changes.

柄搦　其一
右正面寫ノ圖

此モ出方ハ柄取一圖ト同ヿ　双方對顔シ敵ヨリ右足ヲ一足進ムト同時ニ双手ヲ以テ劍ノ柄ヲ圖ノ如クニ取リ引拔ントス　我レ直ニ點線ノ如クニ下ヨリ柄頭ヲ持チ下腹ニ力ヲ入氣合ヲ込テ右足ヲ前ニ進ム同時ニ次圖ニ續ク

柄搦　其二
右斜メヲ寫圖

敵ガ劍ヲ上エ引拔ニ連レテ我レモ右足ヲ蹈出同時

3
Tsuka Karame : **Wrapping up with the Handle of Your Sword**
Sono Ichi : **Step One**
The Attacker is on your right

    This technique begins the same way as step one of Tsuka Tori. Both combatants are facing each other, when the Attacker takes one step forward with his right foot and grabs the handle of your sword with both hands. He then tries to pull it free as shown in the illustration on this page. Respond by immediately grabbing the Tsuka Gashira, pommel, with your right hand from below. This is shown with dotted lines. Fill your lower abdomen with power and take one step towards the Attacker with your right foot. The technique continues in the next illustration.

二下腹ニ力チ込敵
手背部チ圖ノ如ク
ニ掴ミ下リ押エ
テ敵ノ體ヲ崩シ直
ニ右ヘ廻セバ敵手
ガ解ルト共ニ柄逆
ニ成故ニ我カ體ヲ
返ナガラ一足引敵
ノ左手チ放シテ次
圖ニ續ク點線チ見ロ

柄　搦其三

（○此處肝要ナリ）

右斜メヲ寫ス是モ見出
ハ前ノ如ク此處肝要ナ

我レ直ニ右手チ柄
ト共ニ双手ニテ持
左足チ引テ膝チ突

Tsuka Karame : **Wrapping up with the Handle of Your Sword**
Sono Ni : **Step Two**
The Attacker is ahead of you on a right diagonal

The Attacker tries to pull your sword out of the scabbard and up. You respond by stepping forward with your right foot and putting power in your lower abdomen. Push the back of the Attacker's left hand up as shown in the illustration. This will cause the Attacker to become unbalanced. By immediately rotating the handle of your sword to your right, the Attacker's hands will be thrown off. At the same time the handle of your sword will be reversed.

So you then go from facing left to facing right by stepping back with your left foot. This will cause the Attacker's left hand to release. Be sure to study the dotted lines carefully. The technique continues in the next illustration.

キ右足ヲ開キ腰力ダ圍ハレタル心形ヲ見ル内見形ナシ心側ト見正先見ル處ニ此ノ立出ス
キ下開腹入レ除ク敵ヘ残シ附助逸修ルリ
チチ下入腹除敵ヘ散出テラテ立ハ○ヘヲナチシレンチ散手ト宗故流リタヲ見へナルハナ生スチヲ云

70

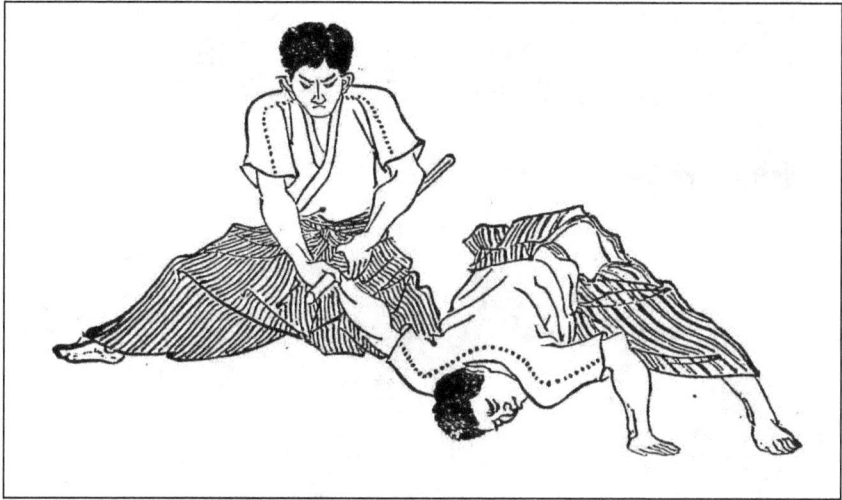

Tsuka Karame : **Wrapping up with the Handle of Your Sword**
Sono San : **Step Three**
The Attacker is in front of you to your left, this is the fundamental point of this technique

With both hands you should immediately grab the Attacker's right hand together with the handle of your sword. Step back with your left foot and plant your left knee on the ground. Extend your right leg out to your right. Put power in your lower abdomen and drop your hips down slowly. You should have the Attacker's right hand wrapped up with both your left and right hands. The Attacker should tap to show he is defeated. Maintain Zanshin while moving back to the starting position.

About this technique :

This is a Tatsumi School technique so Henmi Sosuke Sensei helped adapt it for this book. Henmi Sosuke 逸見 宗助 (1843-1894)

見ル敵ニ足ヲ甲ノ
者如出ニ進者如進見
ハ的前リ出ヲ
下ナ踏ニ合ハ小
腹ニリ甲合太
ニ同前方ハニ刀取
出中方ニ合太
時央同取刀取
ニニ左リ
敵小手左方
ノ大ヲ手ノ
右刀右
手ヲ持ニ距正
首抜チ發離面
ヲキ止スル各
擁チ双ルヲ圖同
ムニ方ト同ニ
ト敵ヲ答ス
ヘル
双
方
共
ニ
圖
右
樣
ニ
同

4
Miai Tori : **Attacked on Sight**
Sono Ichi : **Step One**
The Attacker is on the right armed with a Kodachi, short sword

The Attacker is holding a Kodachi in his left hand and approaches as described in Kyori, or distance, meaning walking towards each other until 90 centimeters apart.

You shout a Kake-goe at him and the Attacker responds with his own Kake-goe. Both combatants approach each other and stop in the center of the training area. The Attacker steps forward with his right foot and moves his right hand to the handle of his Kodachi, indicating he is preparing to draw. You respond by putting power in your lower abdomen, stepping forward with your right foot and seizing the Attacker's right wrist. This is shown in the illustration.

Then immediately step with your left foot towards the Attacker's right side. As you take this second step, maintain your grip on the Attacker's wrist, which will cause his Kodachi to draw out of the scabbard. Your left hand needs to go behind the Attacker and over his left shoulder, so ensure your second step with your left foot, towards his right side, is deep enough to allow you to end up behind the Attacker. The technique continues on the next page.

時ニ右足チ一足蹈込直ニ
左リ足チ又ハ敵ノ右脇ヘ蹈
出ストタン敵ノ剣ヲ拔
連レテ左手ナノ
ヘ附充分ニ敵ノ左肩口
リ込デ次ノ背後ヘ廻
リ込デ次圖ニ續ク

見合取　其二

右斜メ寫圖

甲者ハ敵ノ左肩口ヨリ敵
ノ右襟ヲ摑ミ右手ヲ横ヘ
引左腕ヲ充分ニ延セバ敵
ノ體崩レル故ニ下腹ニ力
チ入レテト敵體ヲ押セ
バ次圖ニ續ク

此畫ニ甲方ノ左腕曲リア
ルハ摸寫ノ都合ニ依テ中
ハ畫書タル處ナリ最モ
畫工ノ働キト云ベシ

（○最モ此處肝要ノ處ナリ）

二十四

74

Miai Tori : **Attacked on Sight**
Sono Ni : **Step Two**
The Attacker is on the right, diagonally in front of you

You reach across the Attacker's left shoulder and seize his right collar. Pull the Attacker's right hand across to your right. If you put sufficient power in your left hand then you will be able to unbalance the Attacker. By continuing to apply pressure, you finish in the state shown in the next illustration. Please understand that this illustration shows the fundamental point of this technique.

You should also note that your left arm is shown slightly bent. This is because the artist is showing the technique in progress.

見合取 其三

甲者ハ充分ニ敵ノ體ヲ崩シテ
我カ左リ足ヲ直ニ後ヘ引キ膝
ヲ突クベシ同時ニ右足ヲ横ヘ
大キク開キ下腹ニ力ヲ入テ腰
ヲ前ヘツキ出スト同時ニ兩手
ヲ押延スベシ敵ノ襟ヲ締メル
ト右手ヲ張ルベシ圖ノ如ク
敵ハ頁ヲ記シテ見合取形　終

○此形ハ(戸田流)ト(喜樂流)ニ
アル形手ノ内ヲ出スモノナリ

片手胸取　其一

双方共ニ距離圖解ノ如クニ進

Miai Tori : **Attacked on Sight**
Sono San : **Step Three**
The Attacker is on the ground in front of you

Having taken the Attacker's balance you next step back with your left leg and drop down onto your left knee. As you drop to your knee extend your right leg out to your right as far as you can. Put power in your lower abdomen and extend your hips forward. At the same time pull both your arms straight. Your left hand will choke the Attacker with the right side of his collar. Your right arm should be extended as shown in the illustration. The Attacker should tap to indicate he is defeated.

This is the end of Attacked on Sight.

About this technique :

This technique is a combination of a version in the Toda and Kiraku schools of Jujutsu.

ミ出敵ヨリ甲者ノ胸襟ヲ両
手ニ集メ左手ニテ摑ミ右足
ナ一足蹈込同時ニ點線ノ如
ク右拳ヲ以テ甲ノ頭上ヘ打
込ナリ　甲者ハ右足ヲ斜メ
ニ開キ同時ニ敵ノ摑ミタル
手首先ヲ圖ノ如ク左リ手
ニテ握リ右手ハ摑ミ居ル襟
下ヲ引上レバ敵ノ指ニスク
ユヱ我レ小指同時ニ敵手ヲ
敵ヨリ打附ル同時ニ敵手ヲ
逆ニ取次　　圖ニ續ク

片手胸取　其二

左正面寫圖

敵ヨリ打込節ニ我カ左手ノ
(コベ)ニテ押ヘ右手添敵ノ
掌ヲ逆ナシテ我両手ニテ圖

78

5
Kata Te Mune Tori :
Defending Against a Single-Handed Chest Grab
Sono Ichi : **Step One**
The Attacker is on the right in front of you

Both combatants approach as shown in Kyori. The Attacker balls both hands into fists and, stepping forward with his left foot, grabs your right chest. He then raises his right hand above his head indicating an impending strike. This is shown by the dotted lines. Respond by stepping forward diagonally to your right with your right foot. At the same time, respond to the hand grabbing your chest as shown in the illustration. Your left hand covers and grips the Attacker's left hand just above the wrist, while your right hand grabs your lapel under the Attacker's right hand. With your right hand pull your collar up, which will pry up the Attacker's hand allowing you to seize his fingers. Starting with his little finger gather all his fingers with your left hand.

Just as the Attacker goes to strike you with his right fist, reverse his wrist in a Gyaku. The technique continues on the next page.

両手ヲ直ニ組合ヘバ我ニ先ンヂテ指ヲ入レテ崩ルレバ力ヲ腹ヘ充分ニ下シ敵ヲ直ニ引延テ押手

左足ヲ最腹ノ慶度ニ此ノ部ヲ見タヂアリスヲ分ケニヨリ曲ゲヲ延充シテヒルチニノ處ニ兩ノ肩ヨリ兩腕甲ニテ塞此處ニアリタル

Kata Te Mune Tori :
Defending Against a Single-Handed Chest Grab
Sono Ni : **Step Two**
The Attacker is on the right in front of you

As the Attacker strikes, push with the Koba, palm of your left hand, as you join your right hand to the Attacker's left wrist. Your hands should look as shown in the illustration. Drop back with your left foot and put power in your lower abdomen. Grip the Attacker's hand with the fingers of both hands and push so the Attacker's arm straightens. With the Attacker now completely off balance the technique moves to the next page.

This illustration depicts the most important aspect of this technique. The illustrator has the arms of the person doing the technique drawn bent, however this is to show how the hands are positioned. When doing the technique ensure the Attacker's arm is sufficiently straightened by adjusting your arms accordingly.

第圖　　正面テ筋立双手ノ

ヲ圖ス双手ニ硬ヲ如ク押ク

三右膝ヲ出シ左膝ヲ付キ

其ノ前ヘ笑シ頭ノ左腹ヲ

三敵ヲキ

取胸ヲ開入ル斜ニ力ヲ締

手ニ手左腹ヲ心ヘテノ下シ

片敵ニ延シ出ヲ肉ノ手形流新木荒ハ此〇シ

附ス、ハ

Kata Te Mune Tori :
Defending Against a Single-Handed Chest Grab
Sono San : **Step Three**
The Attacker is on the right in front of you

Pull the Attacker diagonally to the left and drop down onto your left knee, while your right leg remains upright. This is shown in the illustration. Put power in your lower abdomen and push your hips forward. Push down with both hands, extending your arms completely locking the Attacker's left wrist. The Attacker taps to signal he is defeated and the technique is over. You should maintain Zanzhin, or remain vigilant, until the Attacker returns to the beginning Kamae.
About this technique :
This is an adaptation of a technique from the Araki Shin School.

腕止メ 其一 <small>左リ斜メヲ寫圖</small>

双方進出テ距離ヲ
畳リ乙者ヨリ右足
ヲ一足蹈込ト同時
ニ右拳ヲ以テ甲者
ノ頭部ヘ打込ミ双
方其圖ノ構ノ如シ
甲者ハ下腹ニ力ヲ
入敵ノ顔ヲ見ナガ
ラニ直ニ次圖ヘ綴
ク

6

Ude Tome : **Stopping a Strike**
Sono Ichi : **Step One**
The Attacker is diagonally to the left of you

Both combatants move towards each other judging the distance carefully. The Attacker attacks by stepping forward with his left foot and raising his right fist, intending to strike you in the face. As soon as the Attacker moves, put power in your lower abdomen and stare straight in his eyes. The technique continues on the following page.

腕ヲ打込ミ圖ノ如ク腕ヲ横ニ我レ一文字ニ止メ其ニ右足ヲ右斜メニ進メ右斜ニ受ケシテ止メヽ直ニ後ニ舊圖ヘ敵引

三十

ヲ肝此最クニ矢握首ノ左敵り腰處モ○續圖リヲ腕腕ヨ

86

Ude Tome : **Stopping a Strike**
Sono Ni : **Step Two**
The Attacker is diagonally to the right of you

When the Attacker begins his strike step back diagonally to the right with your right foot and raise you left arm in an Ichimonji, straight line like the Kanji for one 一, and stop the strike. This is shown in the illustration. Immediately after stopping the arm grab the Attacker's right wrist with your left hand. The technique continues on the following page, but note that this is the fundamental point in Stopping a Strike.

腕止メ 其三

左リ斜メテ寫圖

我レハ敵ノ右腕首ヲ取リ同
時ニ敵ノ腕ノ附根ノ處圖ノ
如ク下ヨリ五指ニテ押我兩
腕ヲ張延シ左爪先ヲ外方趾
ヨリ廻シ爪先ニ力ヲ入腰ヲ
ヒチルト同時ニ右足ヲ以テ
敵ノ右足裏ヨリ強ク打拂フ
ベシ　最モ腕ノ引樣ニテ敵
ノ體充分ニ崩レル故兩手ト
足ト三拍子同時ニ倒スベシ
○此處モ肝要ノ處ナリ

Ude Tome : **Stopping a Strike**
Sono San : **Step San**
The Attacker is diagonally to the right of you

Having taken hold of the Attacker's right wrist, immediately strike with Goshi, all five fingers, to Tsuke-ne, the shoulder joint at the armpit. This will cause the Attacker's right arm to extend as shown in the illustration. Put power in both arms and rotate on your left heel so your toes are pointing away from the opponent. Then plant the toes of your left foot on the ground, put all your weight on them and, while twisting your hips, kick hard and sweep the Attacker's right leg from the back with the sole of your right foot.

Since you are pulling on the Attacker's right arm, he is already off balance. You have three points of contact engaged at the same time : your hand on his wrist, your fingers in his armpit and your sweeping kick. This is known as San-byoshi, three points of contact engaged at the same time.   This combination allows you to topple the Attacker. This is the fundamental point of this technique.

手ヲ打テ倒レルベシ最モ
直ニ起上ルニ眞ノ位ノ第
三圖ノ如ニ起上リ甲者ト
對顔シ殘心ヲ附ルベシ
此ニテ腕止メ　形終ル

〇此ハ元起倒流ノ形手ノ
內ヲ奧田先生ト久富先生
ノ正改ヲ猶今回久富先生
ノ修正ヲ受テナス處ナリ

90

腕止メ　其四　見所ノ正面ヨリ写圖

前條ヨリ引續キ敵ノ體ノ充
分ニ亂レタル切右足ヲ高ク
ヨリ敵ノ膝裏ヲ打拂エバ敵
ハ浮足ニナル故ニ拂フト直
ナニ我拂ヒタル足ヲ其處ヘ
蹈止メ左リ足ヲ左ヘ大キク
開キ圖ノ如クノ樣ニナリ敵
ノ起上ル迄下腹ニ力ヲ入兩
股立チヲ持チ殘心ヲ附ル迄
敵ノキョドウチ見込ムベシ
　乙者ハ足裏ヲ拂ハレル故
ニ是非共倒レルコ故ニ左リ

腕止 〆 其四 見所ノ正面ヨリ寫圖

前條ヨリ引續キ敵ノ體ノ充
分ニ亂レタル切右足チ高ク
ヨリ敵ノ膝裏チ打拂ヘバ敵
ハ浮足ニナル故ニ拂フト直
ナニ我拂ヒタル足チ其處へ
蹈止〆左リ足チ左へ大キク
開キ圖ノ如クノ樣ニナリ敵
ノ起上ル迄下腹ニ力チ入兩
股立チチ持チ殘心チ附ル迄
敵ノキヨドウチ見込ムベシ
乙者ハ裏チ拂ハレル故
ニ是非共倒レル「故ニ左リ
手チ打テ倒レルベシ最モ
直ニ起上ルニ眞ノ位ノ第
三圖ノ如ニ起上リ甲者ト
對顏シ殘心チ附ルベシ
此ニテ腕止ノ 形終ル
○此ハ元起倒流ノ形手ノ
内チ奥田先生ト久富先生
ノ正改チ猶今回久富先生
ノ修正チ受テナス處ナリ

Ude Tome : Stopping a Strike
Sono Yon : Step Four
You are on the left and the Attacker has been thrown down in front of you

This illustration continues from where the other left off. With the Attacker's balance taken you have raised your right leg up high and kicked the back of the Attacker's knee with a sweeping strike. This will completely knock the Attacker off his feet. Continue to sweep your right leg back and plant your right foot on the ground. Take a big step out to the left with your left foot. You should end up in the position shown in the illustration.

Keep power in your lower abdomen and hold the sides of your Hakama until the Attacker has recovered and returned to his starting Kamae. Finally, Maintain Zanshin, martial awareness, and observe the Attackers movement and intentions.

Since the back of the Attacker's leg has been swept out from under him he has been thrown down. He should tap the ground with his left hand to indicate he has been defeated. The Attacker should stand up and return to Shin no Kurai #3 stance. He too should lock eyes with you and maintain Zanshin.
This is the end of Stopping a Strike.

About this technique :
This technique was adapted from a Kito School arresting technique. It was modified by Okuda Sensei and Hisatomi Sensei. For this new edition Hisatomi Sensei also made some adjustments to the technique.

襟 投 其一

左斜メ寫圖

此形ハ甲者ヨリ乙者ノ前ヘ
ニ進出テ二尺餘ノ距離ヲ明
甲ヨリ乙ノ襟ヲ両手ニ集テ
右手ニ摑ミ下腹ニ力ヲ入敵
ノ顔ヲ見ナガラ乙者ノ胸部
ヘ押附ルト同時ニ乙者ハ圖
ノ如ク下ヨリ手首ヲ取テ左
リ足ヲ一尺斗リ引下リ甲ノ
顔ヲ見込下腹ニ力ヲ入テ右
拳チナシテ　　直ニ次圖ニ續
クベシ　　此三圖ハ總テ解シ
好クナル處故ニ圖ニ心ヲ止
メテ注意スベシ最モ盡工ノ
苦心ナス處ナリ

三十四

7
Eri Nage : **Collar Throw**
Sono Ichi : **Step One**
The Attacker is diagonally to the left of you

In this technique you approach your Opponent and from about 2 Sun, 60 centimeters, away grab his collar with your right hand. Place power in your lower abdomen and, as you stare him in the face, push his chest. In response your Opponent does as is shown in the illustration, putting power in his lower abdomen, grabbing your right wrist from below and dropping back approximately 1 Shaku, 30 cm, with his left foot as he returns your stare. Your Opponent makes a fist with his right hand. The technique continues on the following page.

This technique has three illustrations which the artist has taken pains to imbibe with the spirit of this technique. Please be sure to study them carefully.

Note : Opponent has been used instead of Attacker since you are making the initial move.

襟投 其二 左リ斜メ寫圖
敵ハ既ニ右拳ヲ以テ打タントスル處ヲ我レヨ
リ右膝ヲ敵ノ股ヘ突キ左足ハ少シ後ヘ延ス右
腕ヲハ充分ニ押突コト圖ノ如ノ樣ニナル故ニ
直ニ我カ右腕ノ下ヲ明ケ敵ノ顔ヲ白限同時ニ
次圖ヘ續クベシ

Eri Nage : **Collar Throw**
Sono Ni : **Step Two**
The Attacker is diagonally to the left of you

As your Opponent raises his fist to strike you drop down onto your right knee between his legs. Allow your left leg to extend behind you and keep your right arm fully extended. You are still pushing as shown in the illustration. You should be glaring at your Opponent with white eyes from below your right arm. The technique continues on the following page.

Note : "Glaring at your Opponent with white eyes…" means to be fully focused on the other person. Your eyes are open wide and show your intensity.

襟投 其三

正面ヨリ寫圖

敵ノ躰ヲ充分ニ押崩テ我カ腕ノ下ナクバリ圖ノ如クノ樣ニ成右膝頭ヲ左足ト共ニ廻リ込ミ左足ヲ左リ後ヘ大キク開キ敵ノ襟ヲ我前ニ引落シ左手ヲ敵ノ左足ニスクイ上ゲルベシ敵ハ我前先ヘ充分ニアチヲ向ニ倒レルナリ起上リハ双方其眞ノ位第三圖ノ如ニテ殘心附テ終乙ハ是非投ラル、故ニ早ク手ヲ打テ頂チシメスベシ

○關口流砂法ノ内形(異揚流)ノ絹澄ノ形手ノ内ヲ修正ス

三十六

Eri Nage : **Collar Throw**
Sono San : **Step Three**
You are in front with your opponent behind you

After pushing until your Opponent has lost his balance rotate your body under your right arm until you end up in the position shown in the illustration. You rotated on your right knee, bringing your left leg around to your Opponent's left side. Then, at the same time, drop your left leg back and pull your Opponent down by his collar in front of you. Use your left hand to scoop up the shin of his left leg. The Opponent should be thrown onto his back.

Finally, both combatants should assume Shin no Kurai #3 stance and maintain Zanshin. Since this is a hard throw, the person receiving the technique should slap the ground to show their defeat quickly.

About this technique :

This technique is adapted from a technique called Kinu Katzuki, Sink Like Silk, from the Sekiguchi and Yoshin Schools of Jujutsu.

摺込 其一

右正面寫圖

雙方進出ルヽ距離圖ト
同シ　敵ヨリ甲者ヲ見
込右拳ヲ振上ル同時ニ
右足ヲ右斜メニ引ナガ
ラ下腹ニ力ヲ入テ甲者
ヘ拳ヲ打込ム同時ニ右
足ヲ前ヘニ踏出スベシ
甲者ハ敵ノ打込樣ヲ見
テ下腹ニ力ヲ入テ只腕
ヲ振上迄ヲ見込テ居テ
直ニ次圖ニ次クベシ

100

8
Tsuri Komi : **Slide In**
Sono Ichi : **Step One**
The Attacker is on the right

Both you and the Attacker walk towards each other as shown in the illustration for Kyori. This means approaching until you are 90 centimeters apart. The Attacker glares at you and raises his right fist. At the same time he steps diagonally backward with his right foot. Placing power in his lower abdomen the Attacker throws a punch downward at you while stepping forward with his right foot.

You have understood the Attacker's intention to strike and have put power in your lower abdomen, however you make no move until the Attacker raises his arm. Continues on the following page.

摺　込　其二

左リ斜メ寫圖

敵拳以テ甲ノ頭上へ打

込故ニ敵ノ蹈出ス右足

ノ外へ左足ヲ我レモ蹈

出ス同時ニ左手ヲ矢筈

ニシテ打落ス腕ニ摺リ

ナガラ敵ノ顎ノ處へ押

當テ充分腕ヲ延スベシ

敵ノ體圖ノ如ク崩レル

故ニ下腹へ力ヲ入テ爪

先ニ充分ニ蹈ミシメテ

直ニ次圖ニ續クベシ

〇最モ此處肝要ノ處ナリ

***Tsuri Komi* : Slide In**
***Sono Ni* : Step Two**
The Attacker is on the right

The Attacker has raised his fist over his head and is preparing to strike. Therefore when the Attacker steps forward with his right foot, you step outside this with your left foot. At the same time form your left hand into a Yahazu, nock of an arrow shape, slide it across the inside of the Attacker's right punch, strike him in the jaw and shove. You should ensure your arm is sufficiently extended. The Attacker's balance should be completely taken as shown in the illustration.

In order to facilitate this ensure you have put power in your lower abdomen and make sure your toes are planted firmly. This is the most important point of Slide In. The technique continues on the following page.

摺込 其三

左リ斜ノ寫圖

我ハ敵ノ體崩レタル故ニ直
右足ヲ一尺バカリ開キ左手
ヲ充分ニ押延セ敵ノ右足ヲ
右外部ヨリ持上テ充分ニ氣
合ヲ込メテ圖ノ如クノ樣ニ
ナル處ニテ次圖ニ續クベシ
但シ此摺込挿畵ノ内第四圖
ノ投ル處甲乙共ニ袴ガ畵工
ノ誤リニテ模樣ノ續キト違
ヒアルハ讀者ニ於テ御察シ
アルコヲ乞

三十九

Tsuri Komi : Slide In
Sono San : Step Three
The Attacker is falling diagonally to your left

Having destroyed the Attacker's balance you then take a short 1 Shaku, 30 centimeter, step to the right while keeping your left arm extended. Gathering your martial energy, scoop the Attacker's right leg up with your right arm from the outside. You should be positioned as shown in the illustration. Illustration 4 for Slide In on the following page has an error by the artist. The Hakama, traditional split skirts, are drawn with different patterns, so the viewer should take that into consideration when viewing it.

ク左手ヲ打テ輕ク倒ル

ベシ起上リニハ眞ノ位

ノ第三圖ノ如ク搆ニテ

残心ヲ附テ摺込形　終

ル

○此ハ（無双流ニ清水流）

手形ノ内ヲ出シタル者

ナリ

敵ノ起上ル迄ハ此ノ搆

ヲ崩スベカラズ残心ヲ

附迄ハ勇氣ヲ附ルベシ

摺込　其四

正面右斜ノ寫図

我レハ敵ノ足ガ浮キタル處ヲ下
ヨリスクヒ上ル様ニナシテ敵ヲ
向ヘ斜メニ下腹ニ力ヲ入テ左手
ハ押延ス右手ハ上グナガラニ向
ヘ投出スベシ
但シ顎ニ手ヲ掛ルハ形ノミ實地
ハ鼻ヲ下ヨリ突上ルコトヲ聞バ其
心得ニテ稽古スベシ敵モ是非躰
ハ亂レ倒レルコニ極ル處故ニ早

107

摺込 其四
正面右斜ノ右図

我レハ敵ノ足ガ浮キタル處チ下
ヨリスクヒ上ル樣ニナシテ敵ナ
向ヘ斜メニ下腹ニ力ヲ入テ左手
ハ押延ス右手ハ上ゲナガラ二向
ヘ投出スベシ
但シ顎ニ手チ掛ルハ形ノミ實地
ハ鼻チ下ヨリ突上ルコヲ聞バ其
心得ニテ稽古スベシ敵セ是非咊
ハ亂レ倒レルコニ極ル處故ニ早

夕左手チ打テ輕ク倒ル
ベシ起上リニハ眞ノ位
ノ第三圖ノ如ク搆ニテ
残心チ附テ摺込形 絡
ル
○此ハ(無双流)(清水流)
手形ノ内チ出シタル者
ナリ
敵ノ起上ル迄ハ此ノ搆
チ崩スベカラズ残心ナ
附迄ハ勇氣チ附ルベレ

Tsuri Komi : **Slide In**
Sono Yon : **Step Four**
The Attacker has been thrown down to the right

You have forced the Attacker onto his back foot, causing his right foot to float off the ground. With your right hand you scoop it up from below. You hurl the Attacker diagonally away from you by putting power in your lower belly and thrusting with your extended left arm and raising his right leg.

Note that when training, the strike to the jaw should be Katachi nomi, or just showing the shape, however in a real situation you would strike up into the bottom of the nose. You should make sure you train this technique with this feeling. The person receiving the technique will be thrown quite hard so he or she should endeavor to roll out of the throw lightly and tap out quickly with their left hand. When recovering from this, return to Shin no Kurai, illustration 3, whilst maintaining Zanshin.

End of Slide In.

About this technique :
This is adapted from the Muso and Kiyomizu Schools of Jujutsu.
You should maintain your fighting spirit and Zanshin in this final
pose until the Attacker has recovered.

敵ノ先　其一
左リ斜メ寫圖

双方進出ルゝ前ニ
同ジ双方共左ヘ一
尺餘斜メニ開クゝ
敵ヨリ右足ヲ前ニ
蹈出ス同時ニ右拳
以テ頭ヘ打込故ニ
我レハ下腹ニ力
ヲ入テ敵ノ様子ヲ
見テ直チニ次圖ヘ
續クベシ

敵ノ先　其二
右斜メ寫圖

四十二

9

Teki no Saki : **Attacker Strikes First**
Sono Ichi : **Step One**
The Attacker is diagonally on the left

You and the Attacker approach each other the same way as in the previous technique. Both of you open your stance by stepping out to the left about 1 Shaku, 30 centimeters, with the left foot. The Attacker begins the technique by raising his right hand above his head and, at the same time, stepping forward with his right foot. You should ready yourself by putting power in your lower abdomen and watching the Attacker carefully.

The technique continues on the following page.

捕形込合ハ此共双リ矢右敵
シ形込合ハ此共双リ圖足ヲ
ヘナナ氣處ニ方ニナリ
讀一打
ク足込
ヘ後拳
シヘナ
〇我ガ
最リ左
モリ手
モリ直ニナ
此處ニナ
敵散横
ノノ
先右受
ノ手止
肝首ヲ同
要ヲ擱時
處ナニ

Teki no Saki : Attacker Strikes First
Sono Ni : Step Two
The Attacker is diagonally to your right

As the Attacker punches, block by raising your left forearm so it is horizontal to the floor. At the same time drop your right leg back one step, then seize the Attacker's right wrist. This is the fundamental point of Attacker Strikes First. Both you and your training partner should study this step carefully and do the technique with sufficient martial spirit. The technique continues on the following page.

敵ノ先 其三
左リ斜メ寫圖

我レ敵ノ右手首ヲ握ヤ
直ニ右足ヨリ腰ヲ敵ノ
腕下ヘ廻リ込右手ヲ以
テ敵ノ二ノ腕ノ處ニ捆
ミ下腹ニ充分ニ力ヲ入
足ヲ柳ニ開キタル樣圖ノ
如ニナシ(エイ)ノ一聲
ト共ニ次圖ヘ續クベシ
○此時ニハ敵ハ早ク投
ル樣ニ心得ヘテ手ヲ
打樣ニナスベシ

四十四

114

Teki no Saki : **Attacker Strikes First**
Sono San : **Step Three**
The Attacker going to be thrown diagonally to the left

You have seized the Attacker's right wrist with your left hand. Next, step forward with your right leg, rotating your body around so your hips are aligned directly under the Attacker's right armpit. Grab the Attacker's right bicep with your right hand. Focus your power in your lower abdomen and plant your legs firmly as shown in the illustration. With a Kake-goe of *Ei!* throw as shown in the following illustration. This throw will be quick so the Attacker should be prepared to tap out.

第四十五図

敵ノ投ゲシ形ニ附ナリ同時ニ敵背負ヲ正面ヨリ捕ヘテ肩下ヘ乱隊心ノ者ヲ修正ナルスヂニ形チ手チンシ形ヲ先キヘ（ニハイスト）落シテ引頃明ヲ神殺活流打チナリ殺活ナリ腕ヲ我引クナチリ左手ニハ此ハ右腕ヨリ○敵如チ前

Teki no Saki : **Attacker Strikes First**
Sono Yon : **Step Four**
The Attacker is thrown down in front of you

This illustration depicts the scene after the throw. You have moved your hips under the Attacker's right armpit, and with a *Ei!* you lower your shoulder as you pull the Attacker forward, throwing him to the ground. This is the same as Se-oi Nage, back carry throw, when doing Randori, free sparring. The Attacker should slap the tatami with his left hand to indicate that he is defeated. Both combatants should maintain Zanshin. This is the end of Attacker Strikes First.

About this technique :

This technique is adapted from a Shinmyo Sakkatsu School arresting method.

帯　引　其一

右横斜メ寛圖

双方共中央ニ進出テ距離ヲ取對顏直ニ双方共左足ヲ斜ニ開キ敵ヨリ右足ヲ一寸出ス同時ニ我帶ノ前チ上ヨリ摑マントスルナリ

甲者ハ敵ノ勣勢ニ眼ヲ附白眼ミナス此ノ構ヘハ最モ柔術家ノ距離ヲ量リ双方ニ氣合ノ乘タル處ナリ　次圖チヨク見ヨ

四十六

118

10
Obi Hiki : **Belt Pull**
Sono Ichi : **Step One**
The Attacker is in front of you diagonally to the right

Both you and the Attacker move to the center of the training space as shown in Kyori, distance. Looking each other in the face you both step out diagonally forward and to the left with the left foot. The Attacker moves his right foot slightly forward and reaches out to slip his hand into the top of your Obi and grab. You detect this motion as you have been watching the Attacker. The Kamae shown in this illustration is the most useful stance a Jujutsu practitioner can adopt. At the same time stare with a "white glare" as you judge the distance to your opponent. Both combatants should be filled with martial vigor.

The technique continues on the next page.

帯　引　其二

右斜メヨリ寫圖

敵ヨリ直ニ右足ヲ蹈出同時
ニ右手ヲ延シテ我前帶ヲ上
ヨリ摑ミ腰ニ力ヲ入テ（ソー
ト）引附ルナリ我ハ下腹ニ力
ヲ入テ腰ヲ張リコタヱナガ
ラ點線ノ如ノ手ニナシ左足
チ敵ノ右横股ノ處ヘ蹈出ス
同時ニ敵腮ニ押當直ニ次圖
ニ續クベシ　最モ此ノ形ハ
左リ手ハ摺込ト同シ丁ナリ

四十七

Obi Hiki : **Belt Pull**
Sono Ni : **Step Two**
The Attacker is in front of you diagonally to the right

In the previous step the Attacker stepped forward with his right foot and, at the same time, reached out with his right hand and seized your Obi from above. Next, focusing his power in his lower abdomen he pulls back with a shout of *Soo-to!* In response you focus your power in your lower abdomen and drop your hips.

Next, as the dotted line shows, move your left arm forward from your side. Step towards the Attacker's right side with you left foot and, at the same time strike and push up his jaw with your left hand. Your left hand is doing a move just like technique #8 Suri-komi, Slide In.

The technique continues in the next illustration.

帯 引 其三 左リ斜メテ篤圖

敵カ我體ヲ力ヲ入テ引附ル故
直ニ腰ヲ張リ下腹ニ力ヲ入ヲ
引レテ同時ニ圖ノ如ク敵ノ右
股ノ横ヘ我ガ左リ足ヲ蹈出シ
同時ニ左手ヲ矢筈ニシテ敵ノ
腮ニ押當同時ニ又右手ヲ以テ
我レヨリモ敵ノ上帯ヲ上ヨリ
摑ミ充分ニ力ヲ込テ腕ヲ延シ
引ト押ト別ニナシテ敵ノ躰ヲ
充分ニ崩シ次圖ニ繼クベシ
〇最モ此ノ處帯引ノ肝要ノ處
ナル故甲者ハ充分氣合ヲ込テ
術ナナスベシ

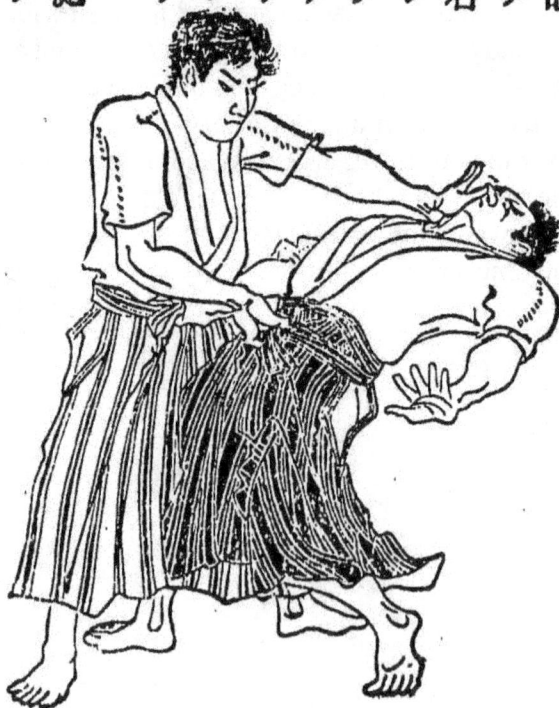

四十八

122

Obi Hiki : **Belt Pull**
Sono San : **Step Three**
The Attacker is being shoved diagonally to the right

This technique begins with the Attacker seizing your belt and pulling powerfully. In response you focus your power in your lower abdomen thereby resisting this pull.

Next, you step forward with your left foot, planting it beside the Attacker's right thigh. At the same time, form your left hand into a Ya-hazu, nock of an arrow shape, and hit and shove the Attacker's jaw. Immediately grab the Attacker's Obi from above with your right hand and grip firmly. Push with your extended left arm as you pull with your right arm. This pushing with one arm and pulling with the other will destroy the Attacker's balance. Pay attention to how you pull the Obi since this is the fundamental point of this technique. Ensure you have sufficient martial spirit in your attack.

The technique continues in the next illustration.

甲者ハ投倒シタレバ體
ニ勢ヲ附下腹ニ力ヲ入
敵ヨリ頂ナシメシ殘心
ヲ附タレバ是ニテ帶引
ノ形〜終ル
○此ハ覔移心頭流ノ形
手ノ內ナリ
最モ此形ハ双方ノ氣合
ヲ充分ニ入テ其術ヲ施
スベシ實地ニテモ隨分
面白キ手ナリ

帯引 其四

右正面ヨリ寫図

我レ前図ノ如ニ充分ニ下腹エ力
チ入（エイ）ト聲ト共ニ敵ヲ斜メ
ニ向ヘ押倒スベシ　乙者ハ又手
ヲ打テアチ向キニ倒レル故ニ早
ク身躰ヲ輕クシテ頁チシメシテ
倒ルベシ起上ル切ハ眞ノ位第三
図ノ如クナナシテ殘心ヲ附テ終
ルベシ

Obi Hiki : **Belt Pull**
Sono San : **Step Four**
The Attacker is in front of you diagonally to the right

As the previous illustration described, you put power in your lower abdomen and together with an *Ei!* shove the Attacker diagonally away from you. The Attacker will fall backwards and then tap out. It is important that the attacker relax his body for the drop and tap out to show he has been defeated. When recovering, the Attacker should return to Shin no Kurai, as shown in illustration #3. Both training partners should maintain Zanshin to the end.

If, during the technique, you are being pulled down, it means the Attacker has won. To ensure this doesn't happen focus your power in your lower abdomen.

This is the end of Belt Pull.

About this technique :
This is technique has been adapted from a Ryoi Shinto School arresting method. It is important that both training partners fill themselves with martial spirit when practicing this technique. When applied in real-life the results are quite interesting.

正面寫圖

此ノ形ハ甲乙連レ達テ
發聲ト共ニ左リ足ヨリ
徐々ト中央迄進出ル成
圖ノ如ク双方共ニ横眼
ニテ敵ノ動勢ヲ見ナリ
最モ行連トアル處ハ總
テ一二尺間ヲ明テ進出
下腹ニ充分ノ力ヲ入テ
両股立ヲ掘リ出ル處ヲ
記シアル故以後出方ハ
此ノ圖ノ如クナリ黑袴
ハ甲　縞袴ハ乙ナリ

128

11
Yuki Zure : **Walking Alongside**
Zu-Sajo : **Upper Left (The Attacker strikes with a right punch so this is in response to an attack from your upper left.)**
Sono Ichi : **Step One**
Both you and the Attacker are shown from the front

In this technique both you and the Attacker are walking in the same direction. The Attacker shouts and begins gradually closing the distance between you, starting with his left foot. As the illustration shows both combatants are glaring at one another from the sides of their eyes, each evaluating the other's movements.

Walking Alongside doesn't really begin until both combatants have closed to a distance of 1 or 2 Shaku, 30~60 centimeters, from one another. You and the Attacker place power in your lower abdomen, as you grip the sides of your Hakama and lift up. As the technique proceeds note that you are in the black Hakama and the Attacker is in the striped Hakama.

行連レ 左上頭 其二
正面寫圖

双方連レ達チ中央
迄ニテ兩足ヲ揃テ
立止ルベシ直ニ乙
者ヨリ圖ノ如ク
甲者ノ顔ヲ見テ直
ニ右足ヲ右斜メ
引開ク同時ニ右拳
ヲ以テ甲ヲ打タン
トニ右足ヲ甲ノ股
ノ處迄蹈込ムベシ
甲ハ下腹ニ力ヲ入
直ニ次圖ニ續クベ
シ

Yuki Zure : Walking Alongside
Zu-Sajo : Upper Left Strike to Your Head (The Attacker strikes with a right punch so this is in response to an attack from your upper left.)
Sono Ni : Step Two
Both you and the Attacker are shown from the front, the Attacker is raising his fist

The Attacker has moved towards you and, having closed the distance to nearly zero, both you and he stop with your legs ready. As soon as you stop the Attacker glares at you and steps out diagonally to the right with his right foot and raises his right hand. This is shown in the illustration.

His next move will be to step forward with his right leg towards your right thigh and punch you in the face. You should put power in your lower abdomen and do as is shown in the following illustration.

行連レ　頭左上　其三

背後寫圖

乙者ハ甲ノ右拳ヲ以テ
打ト見セテ直ニ甲ノ右
肩口ヲ押左手以テ甲ノ
二ノ腕ノ處ヲ掴ミ右足
ヲ以テ甲ヲ既ニ倒サン
トナス甲ハ直ニ體チカ
ワシテ右足ノ趾ニテ入
替リ左リ足ヲ敵ノ兩股
ノ處ヘ蹈込左手ヲ以テ
敵ノ襟ヲ持

Yuki Zure : **Walking Alongside**
Zu-Sajo : **Upper Left Strike to Your Head (The Attacker strikes with a right punch so this is in response to an attack from your Upper left.)**
Sono San : **Step Three**
You are shown from the back

The Attacker has raised his fist and feinted as if he is going to punch you, however he switches his attack and shoves your right shoulder while grabbing your upper arm with his left hand. He then tries to throw you over his leg.

Note: After pushing the Attacker's right shoulder I presume you then grab his left shoulder, but this is not in the text. It is possible the word "right" is a mistake and you push and grab his left shoulder.

You respond by immediately switching your stance. Step back with your right foot and forward with your left. Be sure to step close to the Attacker with your left leg so that you end up deep behind the Attacker, with your left leg extending across the back of both of his thighs. Seize the back of the Attacker's collar with your left hand.

行連 レ 左上頭 其四

正面寫圖

前ヨリ引續キ甲者ハ圖ノ如ク
左手ヲ以テ敵ノ襟ヲ摑ト同時
右膝ヲ突キ右手ヲ以テ敵ノ右
足裏ヲ持敵ノ體ヲ充分ニ崩レ
ル處ヲ以テ直ニ次圖ニ引續ク
ベシ　最モ乙者ハ早ク手ヲ打
テ倒ル心得ナナスベシ
○最モ此形ノ前ト此處ガ肝要
ノ處ナリ

Yuki Zure : Walking Alongside
Zu-Sajo : Upper Left Strike to Your Head (The Attacker strikes with a right punch so this is in response to an attack from your upper left.)
Sono Yon : Step Four
You are shown from the back

This continues directly after the previous illustration. You have gripped the back of the Attacker's collar with your left hand and positioned your left leg behind his thighs. As you drop down onto your right knee, pull down with your left hand and scoop up the Attacker's right leg with your right hand. This will cause the Attacker to completely lose his balance. The technique continues in the following illustration.

The Attacker should slap the ground indicating defeat. You should note that this and the previous illustration represent the fundamental point of this technique.

行連レ 左上頭 其五

正面寫圖

前書ヨリ續ク甲者ハ下腹
ニ力ヲ入テ右手ヲ以テ敵
ノ右足ヲハネ上ル同時ニ
左手ハ下ヘ引落スベシ我
左横合ヘ敵ヲ倒スベシ乙
者ハ早ク左手ヲ打テ頂チ
シメシテ眞ノ位第三圖ノ
如ニナシ双方殘心ヲ附テ
此ノ形 終ル
此ハ殺當流形手ノ內ヲ修
正シテ出シタルナリ

Yuki Zure : **Walking Alongside**
Zu-Sajo : **Upper Left Strike to Your Head (The Attacker strikes with a right punch so this is in response to an attack from your upper left.)**
Sono Go : **Step Five**
You are shown from the back, the Attacker is being thrown behind you

This continues directly after the last illustration. You should put power in your lower abdomen, and lift the Attacker's right leg up and, at the same time, pull down on his collar with your left hand. The Attacker should be thrown down to your left side.

The Attacker should immediately slap the ground to show he is defeated before rising and taking the stance Shin no Kurai, shown in illustration 3 (page 38.) Both combatants should maintain Zanshin, a state of martial readiness until the technique is over. This is the end of Walking Alongside.

About this technique :

This has been adapted from an arresting technique of the Satto School.

行連レ 右突込 其一

正面寫圖

此レハ双方中央迄進出テ立止
ル同時ニ双方右足一尺餘ヲ後
ヘ引對顔シ 乙者ヨリ右拳ヲ
以テ甲者ノ水月(俗ニ水落)ヲ 右足ヲ
蹈出ス同時ニ突込ナリ 甲ハ
敵ガ拳ヲ突出ト直ニ右足ヲ右
ヘ横一文字ニ開下腹ニ力ヲ入
左手ヲ以テ我ガ左脇腹ヲ圍ヒ
手先ニテ敵ノ突出拳ヲ向ヘ押
拂ヒ乍直ニ挿詰ノ
如ク拳ヲ掴ミ右手
ヲ添テ直ニ次圖ニ
續クベシ
○此ノ形大肝要ハ是ナ,
最モ注意スベキ處ト云

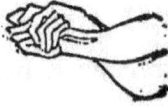

138

12
Yuki Zure : **Walking Alongside**
Migi Tsuki Komi : **Avoiding a Right Punch**
Sono Ichi : **Step One**
You and the Attacker are shown from the front

This technique begins the same as the previous one, with both combatants walking along side each other. As the distance between you closes, you both stop. You and the Attacker both step back with the right foot approximately 1 Shaku, 30 centimeters, while staring each other in the face.

The Attacker steps forward with his right foot and strikes with his right fist to Suigetsu (moon reflected on water,) also known colloquially as Mizu-Ochi (dropping water, the solar plexus.) As soon as you see the Attacker begin his punch, drop back with your right foot so you are perpendicular to the Attacker. This position is called Yoko Ichi Monji, Straight Line to the Side, reflecting the Kanji for the number "one" 一.

Keeping your power in your lower abdomen use your left hand to sweep away the punch as you protect your left side and stomach. Immediately grab the Attacker's hand as shown in the detail illustration, and join your right hand to your left. The technique continues in the next illustration.

You should be aware that proper positioning here is crucial as this is the most important point of this technique.

行違レ　右突込　其二
正面ノ寫圖

敵ノ突込タル拳ヲ摑ム
ト我ガ體ヲ充分ノ
氣合ヲ込ミ敵ノ兩手ニテ
ル我ガ指ヲ敵ノ圖ノ
我テ右足ノ手ヲ敵ノ掌ノ
ニテ同時ニ敵ノ前ニ
蹈出スト如ク上ニ持テア
重ノ如ク左ノ後ニヘア
テ我ガ指ヲ敵ノ前ニ兩手
圖ノ圖ヲ左引クシ左
ゲテ左開クト同時ニハ
キニクヘ
次ニ持レタル乙者ハ
圖ニ繼タル盡直ニ
大圖ニ持レタル盡直ニ心
甲ニ持レタル
早ナク返リ投ラル、心
得ナク返シ投ルベシ
行違レ　右突込　其三
左斜メ寫圖
開ク同時ニ兩手ヲ下ノ
ヘ返シ投ニナスト共

Yuki Zure : **Walking Alongside**
Migi Tsuki Komi : **Avoiding a Right Punch**
Sono Ni : **Step Two**
You are shown from the front

You avoid the Attacker's punch and seize his right hand as shown in the previous illustration. Filling your body with martial power, grip the Attacker's hand with both your hands. Your fingers should be overlapping each other as you step in front of the Attacker with your right foot and, at the same time, raise your hands over your head. Next, take a big step back with your left foot and, at the same time, do as shown in the following illustration. In the next move the Attacker will be thrown quickly to the ground, so he or she should be prepared.

Yuki Zure : **Walking Alongside**
Migi Tsuki Komi : **Avoiding a Right Punch**
Sono San : **Step Three**
You are shown from the front

Maintain a hold on the Attacker's hand as you step back with your left foot. As you continue the throwing motion plant your left knee on the ground while keeping your right knee up. Straighten your arms while putting power in your lower abdomen as you push with your thumbs into the back of the Attacker's hand. The Attacker should slap the ground with his left hand to indicate he is defeated.

Both combatants should maintain Zanshin as they return to Shin no Kurai, as shown in illustration 3. This is the end of Walking Alongside.

行連レ左右腰投其一

正面ヲ寫圖

出方ハ前ニ同甲ハ左

乙ハ右ニ連レ中央迄

進ミ出ル處ニテ乙者

ヨリ甲ノ體ヲ右横ヨ

リ抱込圖ノ如クノ様

ニナス既ニ持上ント

ナス　甲ハ下腹ニ力

ヲ入兩手ヲ下ゲテ充

分ニ腰ヲ張ルベシ

直ニ次圖ニ續ク

144

13
Yuki Zure : **Walking Alongside**
Sa-yu Koshi Nage : **Defending Against a Two-Armed Hip Throw**
Sono Ichi : **Step One**
You and the Attacker are shown from the front

This begins the same as the previous two Walking Alongside techniques. As the Attacker draws near to you, he reaches out with both arms and grabs you from the right with the intention of lifting you up. This is shown in the illustration.

Respond by putting power in your lower abdomen and allow both your arms to droop. Flex your hips and do as shown in the next illustration.

行連レ 左右腰投 其二

正面寫圖

我レ抱込レタル處チ兩肘チ
張リ體チ少シ下ゲ腰チ引テ右
足チ右後ヘ引左足チ少シ前ヘ
蹈出シ敵ノ左腕ノ袖下チ我
ガ左手ニテ下ヨリ掴ミ右手
チ以テ敵ノ右手ノ下ヨリ腰
チ抱込テ引故ニ敵ノ體充分
ニ崩レテ圖ノ如クノ樣ニナ
ル直ニ次圖ニ續ク
○此形ノ肝要ハ此圖ト次圖ニアルナ
リ

Yuki Zure : **Walking Alongside**
Sa-yu Koshi Nage : **Defending Against a Two-Armed Hip Throw**
Sono Ni : **Step Two**
You and the Attacker are shown from the front

The Attacker has seized you with both arms. In response you push your elbows out and drop your hips slightly. Sweep your right foot back and to your left and then step slightly forward with your right foot. Seize the bottom of the Attacker's right sleeve with your left hand while your right hand slips under his left arm and grabs his waist. From that position you pull, completely taking the Attacker's balance. The correct positioning is shown in the illustration. The technique continues on the next page.

Be aware that this step and the next step are the fundamental points of this technique.

Note: The text says, "Sweep your right foot back and to *the right*..." however this probably means "to the right as the viewer sees it."

行連レ　左右腰投 其三
　　　　　　正面寫圖

前書ニ引續キ直ニ我右足ヲ敵
ノ右股ノ先ヘ充分ニ蹈出下腹
ニ我ガ入一寸充分ニ敵體下腹
ノ力ヲ腰ニ引附腰ヲ持上ゲノ
チニ我ハ腰ニテスクヒ上ゲ右
手ノ躰ハ我腰ニ抱込圖ノ如ク右
ニシテ直ニ次ヘ續クハ非是樣
手ナシ、故ニ身ヲ輕クシテ左
投ラル、樣ニ心得ヘ
手ヲ打テ、投ラル、樣ニ心得ヘ
ルベシ

○最モ腰投ト云處ナレバ此處
肝要ナリ

行連レ　左右腰投 其四
　　　　　　左リ斜メ寫圖

甲者ハ敵ノ體ヲ我ガ腰ヲ以テ
ツリ上下腹ニ力ヲ入左手ヲ充
分ニ強ク引附敵ヲ（エイ）ト我

Yuki Zure : **Walking Alongside**
Sa-yu Koshi Nage : **Defending Against a Two-Armed Hip Throw**
Sono San : **Step Three**
You and the Attacker are shown from the front

This illustration immediately follows what was described on the previous page. Step forward with your right foot so that the back of your right thigh is against the front of the Attacker's right thigh. Placing power in your lower abdomen, drop your hips slightly and pull the Attacker's body onto your hip. Lifting him slightly up while keeping a firm grip with your right hand, position yourself as shown in the illustration. You should be carrying the Attacker on your hip and have a firm hold of him with both arms.

The following throw is shown in the next illustration. The Attacker should understand that this will be a powerful throw and therefore should relax his body. The Attacker slaps the ground with his left hand to indicate defeat at the end of the throw.

Since this technique utilizes a Koshi Nage, hip throw, this step is a fundamental point of this technique.

出○テ前ヘ
スハ此残心
形ニ附テ
遊此川ハ乙
ハ流此形ニ投
四形終左タ
方ハ手ル
組テ打
ト頁テ
云テ
形手ヲ
ヲ修シ
正ン
シテ起
テ上上
リ

Yuki Zure : **Walking Alongside**
Sa-yu Koshi Nage : **Defending Against a Two-Armed Hip Throw**
Sono Yon : **Step Four**
You have thrown the Attacker diagonally in front of you

Having loaded the Attacker onto your hip, put power into your lower abdomen and, pulling strongly with you left hand, shout *Ei!* as you throw him forward. The Attacker should slap the ground with his left hand to indicate he is defeated. Both combatants should maintain Zanshin until both have returned to the starting position. About this technique :
This was adapted from the Shibakawa School technique Shi-ho Gumi, Four Direction Attack.

行連レ右衞副其一

正面寫圖

此レモ出方前ニ同
甲者縞袴乙者黒袴
双方中央ニ止同時
ニ乙ヨリ右足ヲ引
直ニ右拳ヲ以テ甲
者ヘ打込同時ニ右
足モ蹈出ス甲者ハ
圖ノ如ク右手ニテ
敵ノ腕ニ摺込止同
時ニ左足ヲ敵ノ右
外股ノ處ヘ蹈ミ直
ニ次圖ヘ續クベシ

14
Yuki Zure : **Walking Alongside**
Migi Kabe Fuku : **Making a Wall Against a Right Punch**
Sono Ichi : **Step One**
You and the Attacker are facing each other

Making a Wall Against a Right Punch begins the same way as the other Walking Alongside techniques. You are in the striped Hakama and the Attacker is in the black Hakama. Both combatants stop when the Attacker has closed the distance sufficiently. The Attacker then steps back with his right leg and then immediately steps toward you to punch you in the face with his right fist. You respond as shown in the illustration. Allow the Attacker's punch to slide off your forearm, as you step with your left foot towards the outside of the Attacker's right thigh. The technique continues in the next illustration.

行連・レ 右體副 其二

正面寫圖

甲者ハ敵ヨリ打込タル右拳ニ
摺レ我右腕ヲ以テ敵ノ右腕ヲ
卷込左手ヲ添テ敵ノ衣紋處ヲ
右手ヲ以テタテニ摑ミ直ニ左
手ヲ敵ノ左肩ヨリ押延シ敵ノ
下襟(左リノ衣ヲ云)ヲ摑ムト直ニ又左足
ヲ引摑ムタル両手ヲ押延シ下
腹ニ力ヲ入テ腰ヲ張ベシ敵ノ
體ハ崩ルルナリ最モ此處ガ肝要
ノ處故ニ此處ニテ眼ヲ附ルベ
シ△圖ニ甲コヽミタレ共下服ヲ前ヘ
出スベシ

Yuki Zure : **Walking Alongside**
Migi Kabe Fuku : **Making a Wall Against a Right Punch**
Sono Ni : **Step Two**
You and the Attacker are facing each other

You have allowed the Attacker's punch to slide across your forearm. Next, with your right arm, wrap up the Attacker's right arm and reach until you can grab the Attacker's Mon-dokoro, family crest (located in between the shoulders.) Pass your left arm over the Attacker's left shoulder and reach for his Shita-eri, lower collar, and grip. Lower collar is another word for his left collar. Immediately step back with your left foot and pull with both hands. You should ensure you have put your power in your lower abdomen.

This will cause the Attacker to completely loose his center of balance. Since this is the fundamental point of this technique, it is important that you study the positioning in this illustration carefully. You should endeavor to thrust your lower hips forward in each step in this technique.

行連レ　右壁副其三
背後宮圖

此レハ前ヨリ引綴キヲ出
ス處ニテ點線ハ左足ヲ蹈
出ス同時ニ敵ノ左リ肩口
ニ手ヲ掛リタル處ヲ記ス
最モ衣紋ノ處ヲ掴ムハ圖
ノ如クナリ右手ヲ以テ敵
ノ腕ヲ卷込ミタル右腕ヲ
解シヨキ爲ニ此處ヲ出ス
ナリ故ニ直チニ次圖ニ續
クベシ　乙ハ下襟ヲ取レ
腕ヲ延バシ喉締ル故ニ顔
ヲ左ノ方ヘ一寸向ベシ

156

Yuki Zure : **Walking Alongside**
Migi Kabe Fuku : **Making a Wall Against a Right Punch**
Sono San : **Step Three**
You are behind the Attacker

Before the technique moves to the next step, this illustration details the movements. The dotted line shows the trajectory your arm takes as you step with your left foot. Your left hand goes over the Attacker's left shoulder. This illustration also shows your right hand gripping the Mon-dokoro, family crest. The purpose of this illustration is to show how your right arm has wrapped up the Attacker's right arm, and grabbed the back of his shirt. The technique continues in the next illustration.

As you pull with your left hand, which is holding the Attacker's collar, it will begin to choke him. He will turn his head to the left to prevent this.

行連レ　右壁刷其四
　　　　正面寫圖

前圖ヨリ續キ我レ下腹ニ力ヲ
入腰ヲ張兩腕ヲ締テ向ヘ押出
セバ敵ノ體亂ル故直ニ左膝ヲ
突キ右足ヲ横ヘ開キ徐々兩腕
ヲ締ルベシ　乙ハ左手ニテ頁
チシメシハ甲ハ兩手ヲ放シ双
方對顔シテ殘心附テ　形終
○此ハ揚心流ノ形手ノ内ヲ出
ス處ナリ

158

Yuki Zure : **Walking Alongside**
Migi Kabe Fuku : **Making a Wall Against a Right Punch**
Sono Yon : **Step Four**
You are behind the Attacker

This is a continuation from the previous two illustrations. Putting power in your lower abdomen and flexing your hips, pull with both arms and it will take the Attacker's balance. His hips will go forward and his upper body will be bent back. In other words, you have completely destroyed his balance.

From there immediately drop down on your left knee as you push your right foot out to the right and gradually increase the pressure with your arms. The Attacker should slap the ground with his left hand to indicate he is defeated. After that you should release your grip and both combatants stand up, maintaining Zanshin until you have returned to the starting position.

About this technique :

This technique was taken from the Torite, or capturing methods, of the Yoshin School.

行連 レ　抽口後其一

右ヨリ寫圖

此形ノ出方ハ甲乙共同處
ヨリ出ル敵カ先ニナリ甲
者ハ後ニテ甲ノ發聲ヲ乙
答ト共ニ二尺餘間アケテ
雙方共ニ下腹ニ力ヲ入テ
進ミ出ルベシ中央迄出テ
乙ハ右ニ甲者ノ顔ヲ見ル
ベシ其樣圖ノ如シ　兩股
立ヲ摑ミテ出ル處ナリ
直ニ止ルヤ右足ヲ右ヘ引
ナガラニ右傘ヲ振上テ甲
者ヲ見込テ　次圖ヘ續ク

行連 レ　後口抽其二

左斜メ寫圖

甲者ハ敵ヨリ右傘ヲ振上

160

15
Yuki Zure : **Walking Alongside**
Ushiro Tori : **Responding to an Attack from Behind**
Sono Ichi : **Step One**
You are on the right with the Attacker behind you

This is a situation where both you and the Attacker start walking from the same spot. The Attacker is in front of you and you are following. You shout at the Attacker and he responds. At this point you have closed to about 2 Shaku, 60 centimeters. Both you and the Attacker should put power in your lower abdomen as you walk forward. When you reach the center of the training area, the Attacker looks back at you. This is shown in the illustration. Both of you should stand in Mata-tachi, the posture of holding the sides of your Hakama up. The Attacker stops and steps out to the right, raising his hand above his head, while looking you in the eyes. The technique continues on the next page.

既ニ同時ニ打込ミ次ニ打込ミ乙ニ受留ムベシ故ニ乙者ハ下腹チ右足チ敵ノ左腹下ニ踏込ミ左腰ニ力チ入レテ乙者ハ甲ニ打込ミ最モ誤チ招キ出スモ右手ニテ甲者チ右手チ右脇ニ拘込ミ前ニ横タフ右手ヲ右手ニテ横一テ以テ

ルニ文字ハ次圖ノ方ダクヨロ、ン此ノ乙者ハ打込ミニ文字ノ圖エ受ケメシ

Yuki Zure : **Walking Alongside**
Ushiro Tori : **Responding to an Attack from Behind**
Sono Ni : **Step Two**
The Attacker is diagonally in front of you

The Attacker, having raised his hand, steps forward with his right foot and strikes down to the top of your head. Respond by putting power in your lower abdomen and launching a counter-strike. Stop the Attacker's hit with a Yoko Ichi-monji, a block with your arm parallel to the floor. Grab the Attacker's waist with your left hand. This is shown on the following page.

The Attacker should strike with all the fingers of his right hand brought together. This is a common mistake make by practitioners so I am noting it here.

行連レ　後ロ捕其三
正面寫圖

甲者ハ敵ノ打込手首ヲ右
手ニテ掴ミ我前ヘ右足ヲ
右ヘ大キク開ト同時ニ敵
ノ體ヲ強ク引ハ乙者ハ亂
ヲ引レル故直ニ左足ヲ乙
者ノ脊後ニ蹈出シ同時ニ
左手ヲ差延シ敵ノ左横腹
ヘ我ガ體ニ引附敵ノ體ヲ
我左股ニ乗掛ケレハ敵ノ
右足ガ浮上ル故右手ヲ以
テ乙ノ右足ヲ押上左手ハ
充分ニ左ヘ廻シ腰ヲ下ケ
テ下腹ニカヲ入直ニ次ヘ

164

Yuki Zure : **Walking Alongside**
Ushiro Tori : **Responding to an Attack from Behind**
Sono San : **Step Three**
This shows you both from the front

The Attacker has punched with his right fist. You stop this and grab his right wrist with your right hand. At this point your right foot is forward. As soon as you grab the Attacker's wrist, take a big step back with your right foot and yank on his arm at the same time. This will disrupt the Attacker's balance. Use this opportunity to step behind him with your left foot and wrap your left arm around his waist, sticking close to his right side.

Pull the Attacker's right leg onto your left leg, which will cause his weight to shift to his left foot and his right foot to come off the ground. Grab the back of his right leg with your right hand. Put power in your lower abdomen and, as you drop your hips apply a twisting pressure to the left with your left hand. The technique continues in the following illustration.

行違レ　後ロ捕其四
正面寫

敵ノ躰崩レタル「故我
ガ體ヲ腰ニ力ヲ入左手
ハ左下へ廻シ　右手ハ
足ヲハチ上ルト同時ニ
（エイヤ）ト腰ト兩手ノ
三拍子揃ヘテ後へ投ル
ベシ　乙ハ早ク手ヲ打
起上リ殘心ヲ附形終
○此ハ各流ヨリ取合シ
テ成タル處ナリ尚今回
久富今泉兩先生ト余ノ
修正ヲナシタリ

Yuki Zure : **Walking Alongside**
Ushiro Tori : **Responding to an Attack from Behind**
Sono Yon : **Step Four**
The Attacker has been thrown behind you

Having broken the Attacker's balance you put power in your hips, sweep you left hand down as you raise you right hand gripping his leg up. At the same time shout a *Ei-ya!* The action of throwing the Attacker behind you with your two hands and hips should be a San-Byo-Shi, or three movements in one. The Attacker should immediately slap the ground, signaling defeat. Both combatants should maintain Zanshin as they return to the starting point.
About this technique :
This technique is a combination of similar methods from all the schools. We consulted with Hisatomi Sensei and he made some corrections.

陽ノ離レ 其一

左正而ヲ寫ス

此形ハ最初ノ出方ト同シ「雙
方其場所ノ中央迄進出テ三尺
距離ヲ量リテ對顏ス　甲者ヨ
リ乙者ヘ仕掛ルベシ　甲者ハ
右五指ヲ固メテ右足ヲ一歩踏
出ス同時ニ乙者ノ兩眼ノ間ヘ
突出スベシ　乙者ハ此ヲ見テ
左足ヲ左ヘ一寸下リ直ニ次圖
ニ引續クベシ
△此ノ袴ハ甲者ノ袴ト乙者ノ
袴トガ間違ヒ畫師ノ誤ニ附島
袴ガ乙者ニテ二圖ヨリハ島ノ
袴ガ乙者ニ相成候間是ヲ正誤
ス

16
Yo no Hanare : Moving Away from Yin
Sono Ichi : Step One
You are on the left with the Attacker in front of you

This technique begins with both combatants moving towards each other. Both of you advance towards the center of the training area while staring each other in the face. Advance until you are 3 Shaku, 90 centimeters apart. In this technique you are attacking first (but the Attacker will still be referred to as the Attacker.) Squeeze the fingers of your right hand together and, stepping forward with your right foot, strike at the point between his eyes. Seeing this attack he avoids by stepping slightly to the left. The technique continues on the following page.

The artist made a mistake when drawing the Hakama of both combatants. In this scene *you* are wearing the "Shima," or black-striped Hakama, and *the Attacker* is wearing the spotted Hakama. However, starting from the next page, the Attacker will have the black-striped Hakama.

Note : The above notation regarding the illustrator's mistake is by the author.
Note: The word *Yin* in "Moving Away from Yin" is half of the Yin-Yang duality.

抱撃ヲ同ニ甲者ハ
附ハシ時ニ者ハ敵陽
ベ敵ニ既ハ右ニ足ノ
シニ甲ニ甲足左ハ右
ガ者參ハ手右
其ノ敵ヘ一以ナ雖
樣右者打ヲ手面モ
次圖參ナ寸ヲ固ル
圖ニ拳込ナ五固テ
ラ抜上引右圖ナ
リ上樣右手打シ
タガ樣手ニ敵
ルナニ拳右ノ
トヲ圖ヲ左眼ニ
以甲ノ左ニ正面
シ者如中正首
ニノ如横中
テ敵
ノ右橫出首
右者出ヲ
ニノヲ如右
ニ龍右様々
ヲ腕左様柔道
ニ様ベ道

Yo no Hanare : **Moving Away from Yin**
Sono Ni : **Step Two**
You are on the left with the Attacker in front of you

You have struck towards the center of the Attacker's eyes with the fingers of your right hand pressed firmly together. The Attacker blocks and sweeps this strike aside with his left hand, which has all five fingers pressed tightly together. At the same time he shifts his right foot slightly backwards and raises his right hand in a fist. The Attacker is now preparing to strike. Your right arm has been swept aside and the Attacker has raised his arm to punch, so you immediately move in and grab the Attacker's body. This is shown in the following illustration.

陽ノ離レ 其三
正面ヨリ寫圖

敵ガ右拳ヲ振テ居
ル所我レ左足ヲ深
ク蹈込左手ヲ以テ
敵ノ體ヲ背ヨリ抱
込右手ハ敵ノ左股
エ押當我右足ハ敵
ノ左足先ニ押止直
ニ次圖ニ續クベシ
乙モ振上タル右
拳ハ甲者ニ抱止ラ
レタル故打下スコ
ナリガタク

Yo no Hanare : **Moving Away from Yin**
Sono San : **Step Three**
You are on the left with the Attacker in front of you

The Attacker has his fist in the air. You respond by stepping in deep with your left foot and wrapping your left arm around his back. Place the palm of your right hand on the Attacker's left thigh and push. Plant your right foot in front of the toes of the Attacker's left foot. The technique continues on the next page. Since you have moved in and grabbed the Attacker he is unable to strike you with his upraised fist.

陰ヲ抱込ミ我右手ヲ敵ニ背ニ掛ケ乙ノ躰ヲ左ニ打チ心ヘ早ク地形シテ投ゲル者ツニニテク地形シテ心ヘ早ク左ノ躰乙

陽ヲ抱込レヨリト同時ニ敵ノ左膝ヲ押ヘ我左手ニテ敵ヲ倒スレ其四畫如クニ斜メ奪盧

Yo no Hanare : **Moving Away from Yin**
Sono Yon : **Step Four**
You are on the right with the Attacker diagonally in front of you

Having wrapped the Attacker up as shown in the previous illustration, twist your body to the left while keeping hold of the Attacker. This technique is colloquially referred to as Randori Yoko Sutemi, Free Sparring Side Sacrifice Throw. As you throw, keep your left hand wrapped around his back and continue pushing with your right hand on the top of the Attacker's left thigh, where it meets the hip. The technique continues on the following page.

The Attacker should relax when being thrown and slap the ground with his left hand to indicate defeat. It is important for the Attacker to recover quickly and face his opponent. This technique is extremely effective in real life so the author has commissioned six illustrations to detail it.

〇此ハ元扱心流ノ手ノ内ヲ修正ヲナシテ出シタル者ナリ
最モ此ノ柔術ノ形ハ筆ニ蓋不蓋ニ見セ不實ニ此度形ハ蓋工ノ動ト云ベ
シ　猶久富大先生ノ望ミニハ我レ
覺ヘタル亂捕ノ新案妙手ヲ表裏ノ
形ニナシ此ノ本ニ引續キ出版ヲ余
ニ談話有之尙蓋師ニ實地要所ノ圖
蓋ヲ摸寫ヲナサシメ拳法同樣ニ何
人ニモ解シ好ク又何流議チトハズ
極意ノ妙手ナレバ出版ノ日ヲ待尙
賞セラレンコヲ望ム

陽ノ離レ　其五　正面窩圖

此ノ處ハ甲者ニ投出サレタル處圖ナリ
乙者ハ左手ヲ打チ倒レタルナリ　乙者ハ
直ニ起上ルニモ左リ膝ヲ突キ右膝ヲ立甲
者ト對顔シ殘心ヲ附テ　眞ノ位ノ第三圖
ノ構ヘヲナスベシ　最モ此處ハ乙者ガ先
ヘ盡ガ出テアルハ甲者ニ投ラレタル處ヲ
見ヨ・
○甲者ハ敵ノ體ヲ抱込我身ヲ返リ兩手ニ
押ハヲ左向ヘ敵ヲ右足ヲ添ヘテホウリ投
ルナリ我レハ一時ハ寢コロブトモ直ニ此
ノ圖ノ如クノ構ヘニテ殘心ヲ附此形終ル
直ニ雙方共ニ元ノ座ニ附禮ナナシテ十六
乎ノ拳法ノ形　終ルナリ

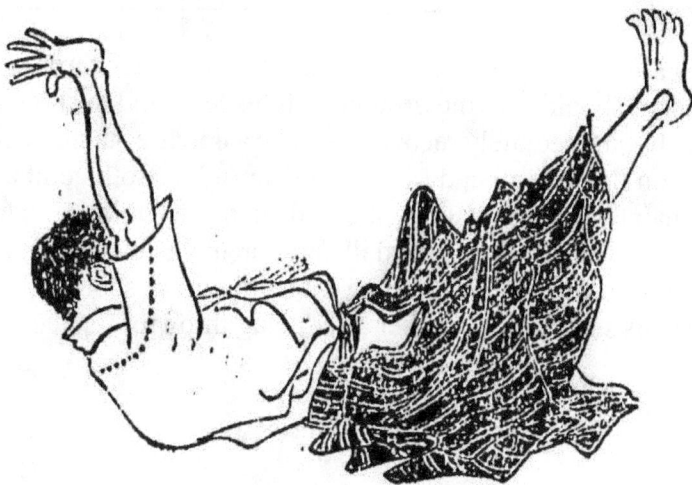

Yo no Hanare : **Moving Away from Yin**
Sono Go : **Step Five**
Scene from the front

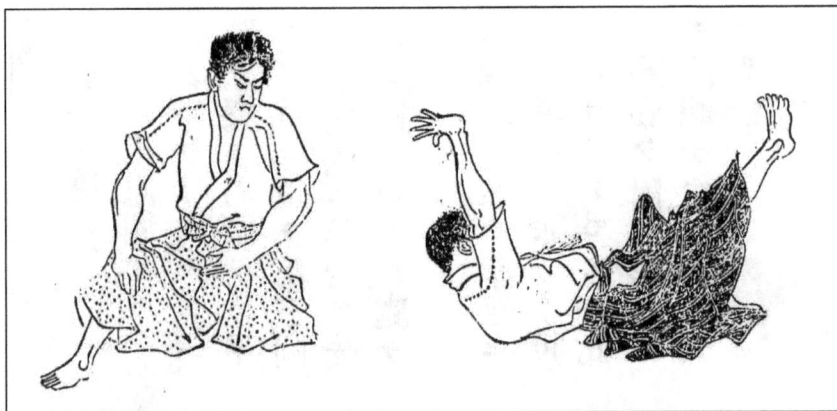

This illustration shows the Attacker after you have thrown him. The attacker should slap the ground with his left hand to show he is defeated, but immediately recover and take a defense stance with his left knee on the ground and his right leg upright. Both combatants should maintain Zanshin until the end of the technique. This is shown in Shin no Kurai, the third illustration in this book. Typically the illustration of your positioning after the throw should be first, however this shows the Attacker after being thrown.

Review of the Final Technique

In this technique you grab the Attacker and, pushing with both hands, twist your body to the left. In the process of throwing the Attacker over by using your right foot to trip him, your body will end up flat on the ground. You should immediately recover and return to Kamae, as shown on the following page, all whist maintaining Zanshin. At the end of the technique both combatants should return to the starting position and do a Rei, bow of mutual respect. Thus this ends the 16 Techniques of the Kenpo.
About this technique :
This was adapted from a Te no Uchi technique from the Kyushin School.

Traditionally Jujutsu techniques were not things that were written down in books or shown with pictures but rather were things that were observed first-hand. These should be thought of as illustrations of movement.
It was the desire of the great Hisatomi Sensei that the reader gain a sense of how Randori should be done. He feels a new approach to understanding the Hyo-ri, Outer and Inner, mysteries of the techniques is necessary. Thus we are working on a follow up volume to this, and the illustrator is currently observing training at various Dojo in order to capture the movements in his drawings. His goal is to make them as easily accessible as we have done with Kenpo. As a final note, no matter what school of martial arts you may belong to, if you have a mysterious, effective technique we would be happy to include such a method in our next publication.

早繩捕縛圖解注意

總テ早繩ニハ種々アリ其數多クシテ大略實地必用ノ處ヲ記シ尙捕繩モ
左ノ如ク流儀モアリ其種類ハ實ニ數ヘ難キ故ニ四五種肝要ノ處ヲ記ス處
最モ是ニ記載アル處ハ此ノ拳法ノ形ト同時ノ頃ニ警視廳ニ於テ各先生
試驗ノ上揷畫ヲ以テ記載セシ者ナリ　早繩ノ心得注意ナス可キ者ハ
警官ノ外ハ猥ニ人ヲ捕縛スベカラス我レニ武術ノ心得アリトシテ野山
ニ行不意盜賊ニ出合組臥テモ必本繩ヲ掛ルベカラズ假ニ繩シ置テ其近
邊ノ警察署又巡査ノ交番處ニ訴ヘルベシ又宅ニ夜盜忍入タルモ前ニ同
○捕繩製造法ハ麻苧ヲ極柔ラカク打三ツ緒ヲ綯ミ細キ方ヲ好トス最モ
繩ニ定法アリ早繩ニハ通常三尋半、五尋半、七尋半ヲ製スルナリ本繩ハ
十一尋以上モアリ繩ノ色ニ註解アレ共別ニ必用無之爲ニ略ス又余ガ著
ス柔術劍棒圖解及武道圖解秘訣ニ本繩圖解アリ故ニ現今必用ノ?ニ記ス
次回ニ壹章毎ニ揷畫ヲ以テ詳細ニ記載スレバ其說明ヲ?讀セヨ

Arresting Rope Techniques

Haya Nawa Hojo Zukai Chu-i
Illustrated Explanation of Fast Tie Arresting Rope

There are many different varieties of Haya Nawa, Fast Tie. I have consolidated these methods, listing only the ones that will be most practical in a real situation. As I mentioned, there are many different schools of Hojo, or binding, and each has its own philosophy. There are so many varieties it would be hard even to name them all, so I have limited myself to four or five fundamental techniques. These techniques should be learned in tandem with the Kenpo techniques.

The Police Department consulted with various martial arts instructors and, after getting approval for each chapter, commissioned illustrations to be made. Understand that these Hojo binding techniques are not to be used to tie people in a perverted way but only employed by officers of the law. The reason we, as martial artists, feel this is important is that you may one day find yourself out in the wilderness and come upon a thief or brigand. In such a situation you will not have the luxury of applying Hon Nawa, a full-fledged restraining tie, rather you will have to resort to something simpler in order to transport the villain to the police station or local constabulary. In addition, if you capture a robber sneaking into your home at night, you will have to respond the same way.

The way to make the best Hojo rope is to pound Asa-o fibers until they are very soft. Braid three pieces of rope together remembering that a thinner rope is better. There is a pre-determined length to Hojo rope. Haya Nawa, Fast Tie rope, typically comes in three sizes, 3.5 fathoms (5.3 meters), 5.5 fathoms (8.3 meters) and 7.5 fathoms (11.3 meters.) For Hon Nawa, a Main Tie rope, the ties are more involved so usually an 11 fathoms (16.5 meter) length of rope is used.

There is a lot of information about the color of the rope used, however this is not of primary importance and will be abbreviated in this book. For details regarding using colored ropes please refer to the author Inoguchi's books *An Illustrated Guide to the Secrets of Jujutsu, Kenjutsu and Bojutsu* as well as *An Illustrated Guide to the*

Fundamentals of Military Strategy. Thus this volume will only focus on the fundamentals. The following techniques will each be presented with an illustration and a detailed explanation. You should read each entry carefully.

Notes on the lengths used and types of rope :

Hiro/ Shin : Fathom
尋 Hiro or Shin, can be translated as fathom since it is used to measure the length of rope as well as the depth of water. The length of one Japanese fathom is not precise, something between 5 and 6 shaku 1.5~1.8 meters. For simplicity's sake I will refer to 1 fathom as 1.5 meters.

Shaku 尺 :
One Shaku is 30.3 centimeters, however at different times in the Edo era the length varied.

Sun 寸 :
One Sun (pronounced "soon") 1/10th of a Shaku, 3.03 centimeters.

Asa-o 麻苧 : Mixed fiber rope
Asa-o is a type of rope made from hemp fibers. The Kanji 麻 refers to hemp and the Kanji 苧 refers to ramie からむし the product of mixing these two fibers is Asa-o. Ramie is one of the strongest natural fibers. It exhibits even greater strength when wet. Ramie fiber is known especially for its ability to hold shape, reduce unravelling, and introduce a silky luster to the fabric appearance. It is not as durable as other fibers, so is usually used as a blend with other fibers such as cotton or wool.

The two books mentioned by the author are :

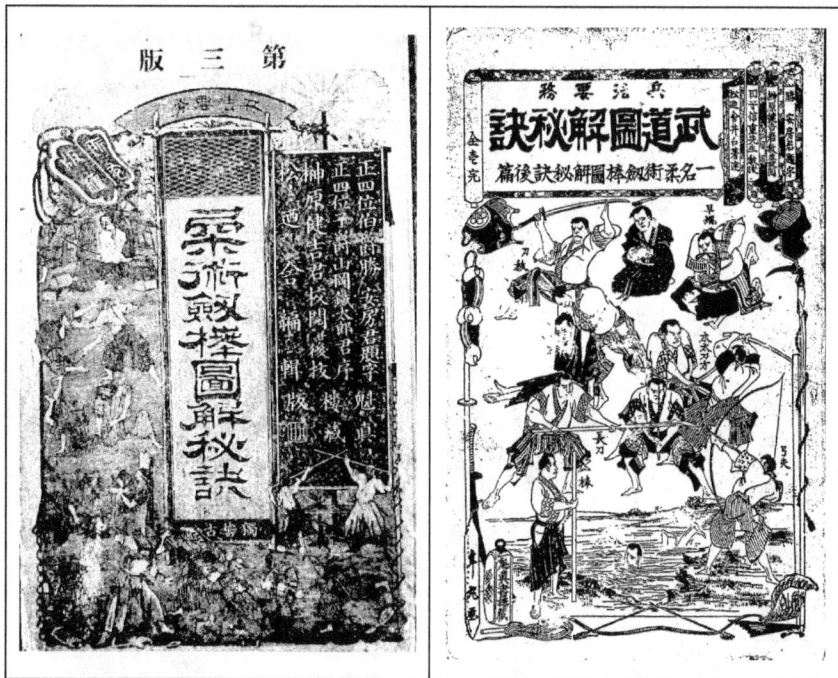

Left :

● 兵法要務武：柔術剣棒図解秘訣
  Heihoyomu : Jujutsu Kenbo Zukai Hiketsu
  Fundamentals of Military Strategy : An Illustrated Guide to the Secrets of Jujutsu, Kenjutsu and Bojutsu
  By Inoguchi Matsunosuke 井口松之助
  Published 1887

Right :

● 兵法要務武道図解秘訣 ：一名・柔術剣棒図解秘訣後篇
  Heihoyomu Budo Zukai Hiketsu : Ichimei : Jujutsu Kenbo Zukai Hiketsu Gohen
  An Illustrated Guide to the Fundamentals of Military Strategy : A Follow up Volume to Secrets of Jujutsu, Kenjutsu and Bojutsu
  By Inoguchi Matsunosuke 井口松之助
  Published 1890

◎釣繩圖解 作寫圖

鉤繩ヲ製造スルニハ第一鉤ハ双金ヲ以テ挿畫如クニ造リ寸法ハ一寸五

分位ヲ善トス盗捕縛ノ切亂暴ナセバ耳ニ掛ルコモアリ襟ニ掛タレバ

手ノ擧ヲ抱合ニテ縛ルベシ

手節ト節トノ間ヲ凹タル

鉤繩其一

處ヲ二卷迴シテ左右ノ手開キ割リテ直ニ横

二玉卷迴シ襟下八寸ノ處ニテ是ヲ垣根結ビニシテ結止ルナリ最モ罪人

ノ帶ヲ好ク締メテクベシ萬一鉤繩永クシテ殘レバ腰ニ卷附テモ善トス

釣繩圖解其二

Kagi Nawa Zukai : Illustration of the Arresting Rope With Hook

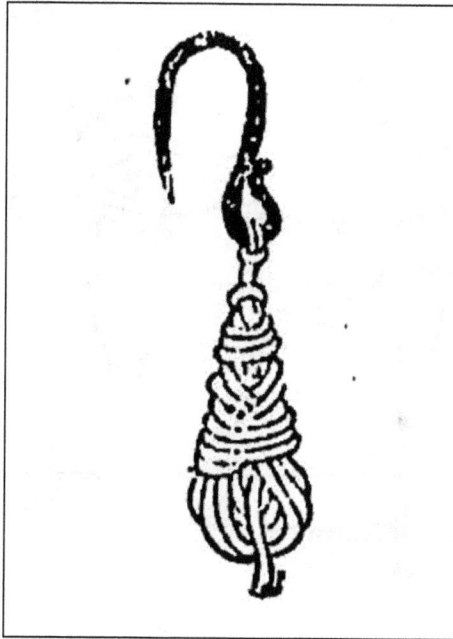

Illustration of Kagi Nawa #1

When making a Kagi Nawa, Rope with a Hook, the first thing to make is the hook. This should be made with knife steel and should look like the illustration. The hook should be about 1 Sun & 5 Bun, 10.5 centimeters long (I presume this means from the center of the hook.) Typically the hook is attached to the collar, however if a thief resists or becomes violent in the course of applying the Hojo, arresting rope, then the hook can be hooked on his ear.

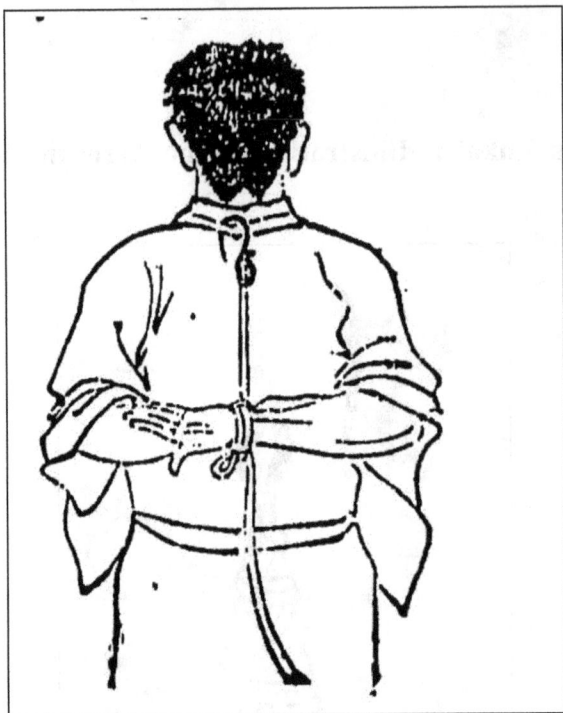

Illustration of Kagi Nawa #2
View from the back

When tying, make the prisoner join the palms of his hands together. Wrap the rope twice around the wrists where a depression forms. Then spread the left and right hands apart and wrap the rope twice horizontally around the gap between the two hands. The final knot in this tie should be a Kakine, fence-knot, approximately 8 Sun, 24 centimeters from the back collar. It is a common practice to tie the final knot to the criminal's Obi, or belt. If additional precautions are desired, or the prisoner will be tied for a long time, the rope can be tied around the prisoner's waist.

Note: The illustrations on the following page show how to tie the Kakin, fence-knot. The illustrations are from Fujita Seiko's book *Samurai Bondage* published in 1964.

Musubi Kata 結び方 Knots

Otoko Musubi: Men's Knot Other names for this knot include; Kaine Musubi (Fence Knot,) Ibo Musubi (Wart Knot,) and others.

How to tie a Men's Knot/ Gate Knot	How to tie a Men's Knot

結び方 一 1

二 2

三 3

男結び

別名 をろ結び もろ結び 垣根結び 襷結び 疣結び 庵結び 技折結び 蠅頭

いがら結び ろ結び

結び方 一 1

二 2

三 3

捕繩圖解

早繩捕繩モ捕押ヘハ挿齒ノ如ク成盜ヲ捕
ヘルニハ盜賊ノ右手先ヲ摑ミ腕ノ節ニ左
手ヲ添ヘ盜ヲ組臥テ其節ニハ我右腕ニ挿
齒ノ環ノ處ヲハメ置テ盜ノ右手ニ我手ヨ

ノ繩先ヲ環
シテ通シタ
ヲル處

リ盜ノ手ニ運シ左膝ヲ盜ノ袴腰ニ押當右
手ヲ曲ゲ盜ノ左肩口ヨリ腮ニ引掛ヌ様ニ
咽ヘ廻シ次圖ノ如クニナスベシ

188

Various Schools
Hojo Zukai : **Illustrated Guide to Arresting Rope**
Sono Ichi : **Illustration 1**
The criminal is diagonally to your left

Illustration showing loop on the end of an arresting rope
(The description has been cut off at the top, however it is typically
called a "snake's head.")

Haya Nawa Hojo, Fast Tie Arresting Rope, is done as shown in
the illustration. Grab the thief's right hand and fingers with your
right hand. Take hold of the thief's right elbow with your left hand
and push him face down onto the ground. Next, as the illustration
shows, transfer the loop of your Hojo rope from your right wrist to
the thief's right wrist. Plant your left knee in the small of the thief's
back on the Hakama-goshi, back of the Hakama. Next, bend the
thief's right arm and pull the rope over his left shoulder and under
his jaw. Then, bring the rope around his right shoulder. The
technique continues on the following page.

捕繩圖解 其一 左ノ正面寫圖

左ヨリ右ヘ廻シ盜ノ咽ヘ掛ケテ萬一盜亂
暴ナナス時ニハ右足ヲ盜ノ右ニノ腕ニ蹈
附左手ヲ折リ曲ゲ左手首ニ二卷繩ヲ附テ
直ニ次插譜ノ如クニ結ビ附テ又第三第四
第五圖迄ニテ終ル最モ此繩モ襟元七八寸
下ノ處ニテ結ブ

○此形ハ元武藤迪夫先生及今泉八郎先生
大竹先生其他各流ノ試驗ノ上ニテナシタ
ル處ナリ此圖ハ今泉先生余ヲ繩シタル處
ヲ安達吟光摸寫スル圖ナリ

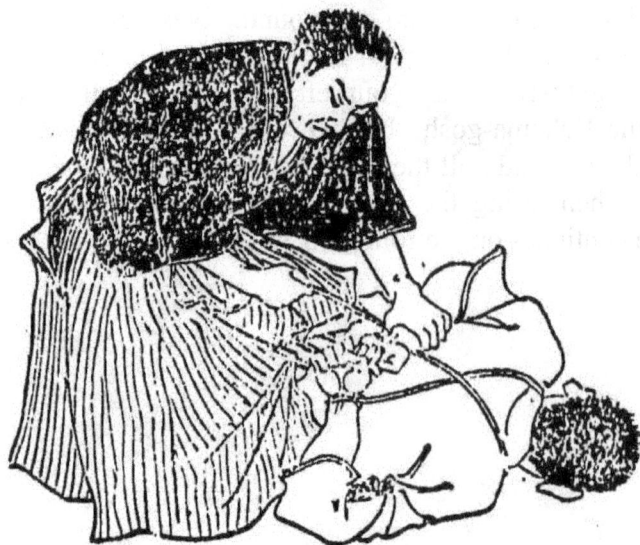

Various Schools
Hojo Zukai : **Illustrated Guide to Arresting Rope**
Sono Ni : **Illustration 2**
The criminal in front of you, slightly to the left

You have passed the rope across the thief's left shoulder, under his neck and over his right shoulder. If the thief becomes violent step on his right bicep with your right foot. Next, bend his left elbow up and wrap his left wrist twice. This is shown in the illustration. In addition, illustrations 3, 4 & 5 show the end of the technique. The arms should be tied 7 or 8 Sun, 21~24 centimeters, below the collar.

About this technique :

This technique was adapted from various schools of martial arts by Muto Michio Sensei, Imaizumi Hachiro Sensei and O-take Sensei amongst others. This illustration by Ando Gin captures Imaizumi Sensei applying the technique.

(I suppose this means that this is a portrait of Imaizumi Sensei.)

捕縄圖解 其三四五 左斜メ寫圖

第三圖解第二圖ニ手首ニ卷廻シ直ニ手元ノ第四ノ如ク左手ノ方ヘ廻シテ引締直ニ第四圖ノ如ク二本同一ニ又通シテ兩手首ニヘ寄ルテ直ニ

第 三 圖

Illustration 3

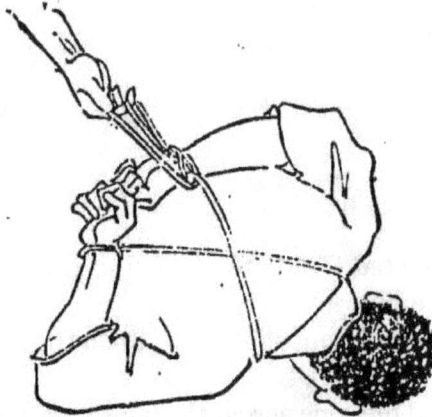

第 四 圖

Illustration 4

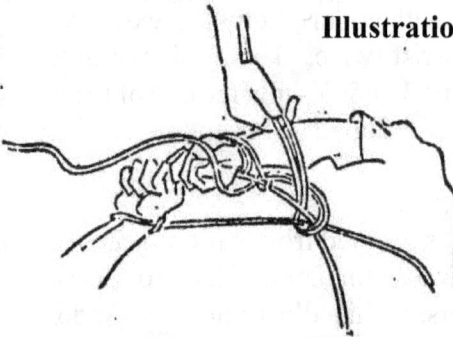

第 五 圖

Illustration 5

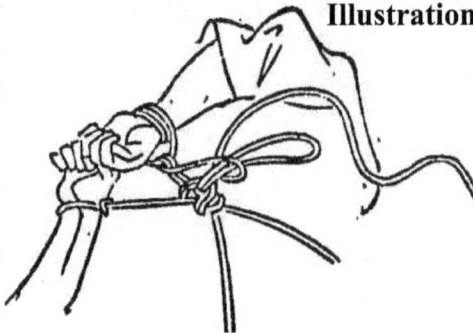

第五圖ノ如クニ尚好ク締メテ手先ヲ重子テ縛ルコヲウルベシ最モ形ハ

四五ノ處ニ頭無トモ縄ノ處ヲ肝要故此處ヲ記スナリ

Various Schools
Hojo Zukai : **Illustrated Guide to Arresting Rope**
Sono San : **Number 3**
The criminal is diagonally to your left

Illustration 3 shows the state of the tie immediately after illustration 2. Illustration 4 shows, after tying off the left wrist, how you double the remaining rope to tie an additional knot. Pass the rope under where the two ropes cross the back and make one loop. The rope should feed towards the thief's left hand. After doing as illustration 4 shows, pass the doubled rope around and through again.

Pull the loop tight so it pulls towards the hands. Illustration 5 shows how a secure knot should be tied just above the hands. The final tie shown in illustrations 4 & 5 is the obvious final step, however as it is a fundamental point of this technique and needs to be remembered, illustrations have been included.

早縄圖解 其一 正面寫圖

此捕繩ハ二筋ニテ手首ニ圖ノ如クニ掛盗
ヲ倒シ其上ニ馬乗ニ股ガリ暴動ナナス時
ハ右耳下ノ處ヲ俗ニ獨古ト云處ヲ右拇指
ノ先ニテ強ク押附テ左手ヲ曲ゲテ捕縛ス
ベシ直ニ第二圖ノ如クニウツルベシ
○最モ此ノ形ハ大原正信先生關口流ノ捕
繩ヲ講修シテ出シタル者ナリ
△尚盗亂暴ナナス時ハ前ノ如クニ二ノ腕ニ
足ヲ掛ケ蹈附ルモヨシトス

Sekiguchi School
*Haya Nawa Zuka*i : **Illustrated Guide to Fast Tie Arresting Rope**
Sono Ichi : **Illustration 1**
The thief is being held down in front of you

This Hojo, arresting rope, technique uses a doubled rope to tie off the thief's right wrist. This is shown in the illustration below. After taking the thief down mount him like a horse. If the thief becomes violent use your right thumb to press hard into the spot below the ear known colloquially as Dokko. (The Kanji 獨古 mean "alone" and "old." This point probably refers to the Buddhist symbolic implement Vajra. The Vajra symbolizes both indestructibility and an irresistible force.)

Use this distraction to bend his left arm and tie him up. The technique continues in illustration 2.

About this technique :
This Hojo technique originates from the Sekiguchi School and was introduced by Ohara Masanobu Sensei. (The text makes no mention of if this is a portrait of Ohara Sensei, though it seems likely.)
If you encounter a thief that is particularly violent you can also step on his bicep as was shown in the previous technique.

解圖二第

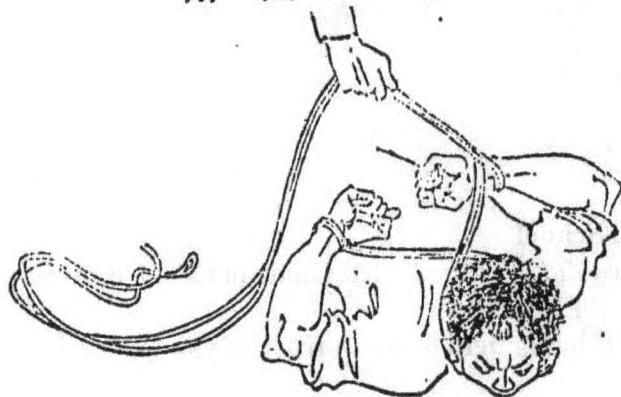

早縄圖解 其二。寫圖ノ處ハ前ニ同

此ノ如クニ左リ手首ニ
卷附レバ第三圖ノ如ク
ニ手ヲ引締メテ襟下ハ
寸ノ處ニテ縛リ抱起ス
ベシ最モ此形ニ於テハ
一筋ノ縄ニテ縛リ兼タ
ル時ニハ二重ニテ縛ル
處ヲ出スナリ第三圖ノ
如クニテ終ル

解圖三第

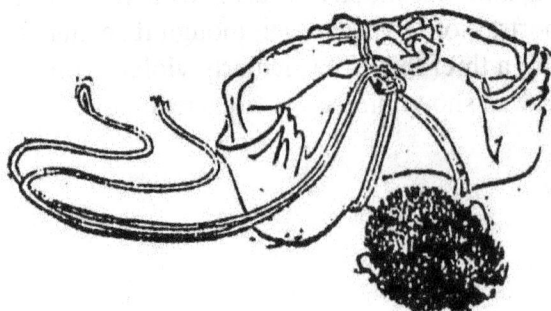

Sekiguchi School
Haya Nawa Zukai : **Illustrated Guide to Fast Tie Arresting Rope**
Sono Ni San : **Illustrations 2 & 3**
Image is showing the same view as before (The thief is being held down in front of you)

The rope goes around the thief's left wrist as shown in illustration 2. Next the hands are pulled together and tied off about 8 Sun, 24 centimeters, below the collar. Having completed the arresting tie, raise the thief to his feet.

Originally this technique was done with only a single strand of arresting rope, however when possible the doubled rope version shown here is more effective. The technique ends as shown in illustration 3.

捕繩圖解 其一 右正面寫圖

此捕繩ノ圖解ハ押ヘ方ハ各前ニ同ジフ
直ニ盜ナウツ向ニ倒シ左リ足ニテ盜ノ
二ノ腕ヲ强ク蹈附テ前ノ如ク左肩口ヨ
リ咽ヘ廻シテ引掛ケ左手首ニ二卷廻シ
テ結ビ附ルベシ最モ此レハ酒ニ醉ヒ亂
暴ナス者ヲ縛スルニハ第二圖ニ及ヒ第
三圖ノ如クニナスベシ
此形ハ故逸見宗助先生家傳立見流捕繩
形ナリ

Henmi School
Hojo Zukai : **Illustrated Guide to Arresting Rope**
Sono Ichi : **Illustration 1**
You are holding the thief in front of you

This illustrated example of how to apply the Hojo arresting rope begins in the same way as the previous examples. You have taken the thief to the ground and have stepped hard on his upper arm with your left foot. As before, after you tie the rope around the thief's right wrist wrap it over his left shoulder, around his throat, over his right shoulder and down towards his left side. Wrap the Hojo rope around his left wrist twice and tie a knot.

This method is ideal for subduing intoxicated and violent criminals. Illustrations 2 & 3 will show additional details.

About this technique :
This is a direct transmission from the Tatsumi School as taught by the recently departed Henmi Shusuke Sensei (1843 – 1894.)

Illustration 2

第二圖解

處ヲ記シヌ

捕繩圖解　其二　竅ノ處ハ前ニ同

第二圖ノ如クニナシ締縛タレバ其儘ナシ置ベシ萬一亂暴ナナス時ハ直

ニ縄ノ殘リヲ以テ
第三圖ノ如ニ左右
ノ内足ノ拇指一本
ニ結附ルベシ倒シ
置バ醉ノサメタル
時ハ解クベキナリ
警視廳ニ於テ各流
先生ノ良法ヲ出ス

Illustration 3

第三圖解

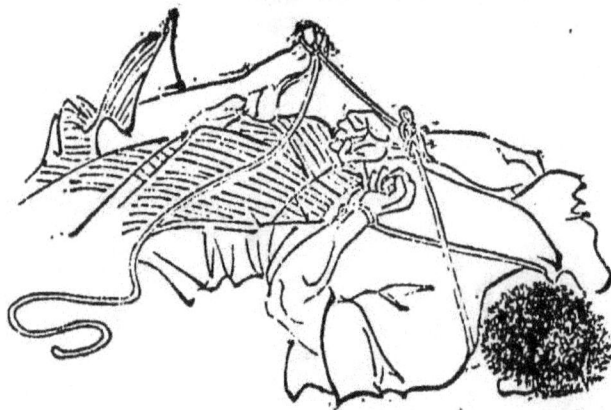

Henmi School
Hojo Zukai : **Illustrated Guide to Arresting Rope**
Sono Ni San : **Illustration 2 & 3**
You are holding the thief in front of you

This method ends as shown in illustration 2. However, if the drunkard becomes violent then you should immediately use the remaining rope to do as shown in illustration 3. Take hold of either his right or left foot and tie off one toe and leave him on the ground. When the drunkard has sobered up, you can untie his foot. This is a police department technique that has been judged effective by the heads of all the schools of martial arts, so no further instruction is necessary.

Illustration 1

第一圖

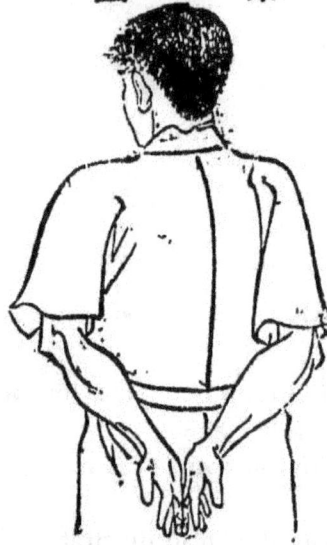

Illustration 1

Illustration 2

第二圖

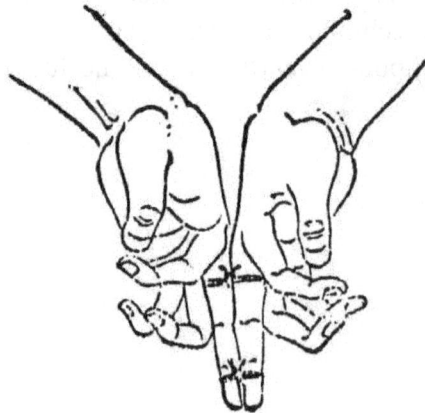

Illustration 2

早縄圖解　背寫圖

此ノ早縄ノ製造方ハ麻苧極上等又絹絲ノ太キヲ用ヒ又元結紙緒ニテモ

多勢ニテ縄ノ間ニアワザル時ニハ圖ノ如クニ綯ルベシ最モ第一圖ノ如

ク手背ヲ合シ第二圖ノ如ク二中指ノ附根ト先節ノ處チ堅ク結ブベシ最

セモ博徒又一揆ナゾニハ縄ノ不足ノ時ニ前書ノ如ク五寸七寸一尺迄チ用

ヒルナリ○此形ハ水野流ニテ國重重信先生ノ講修スル處ナリ

Mizuno School
*Haya Nawa Zuka*i : **Illustrated Explanation of Fast Tie**
Sono Ichi Ni : **Illustrations 1 & 2**

The string used in this Haya Nawa, fast tie, technique is either high-quality Asa-o 麻苧 (a type of rope made from hemp fibers,) thick silk thread, Moto Yui (short pieces of string used to tie Samurai topknots, or Kami-o (general purpose paper string.) This method can be used when you have to restrain large numbers of people and don't have enough rope. The illustrations show how this is done.

Illustration 1 shows the criminal with his hands behind his back. Illustration 2 shows two ties. First make a secure tie at the base of the middle fingers and then make another secure tie at the first joint of those same fingers.

This technique was originally used to arrest large groups of gamblers or rioters. In such cases, since rope may run short, this method is used with 5 Sun (15 centimeter,) 7 Sun (21 centimeter) or 1 Shaku (30 centimeter) pieces of string.

About this technique :
This is a Mizuno School technique as taught by Kunishige Shigenobu Sensei.

Note : I was unable to find any information regarding Kujishige Shigenobu Sensei.

早縄圖解 背寫圖

此ノ形ハ兩拇指ヲ脊ト脊ヲ合シ充分ニ結附直ニ襟ニ通シ圖ノ如クナシ

結附七寸繩ヲ用ヒタル處ナリ最モ急塲ニテ繩ノ手廻リ兼ル時ハ前條ノ

五寸繩ニテ好シトス其役等者ハ常ニ絹糸ノ太キ

處ヲ人血ヲ以テ五寸七寸一尺等ヲモ製造シテ置

キ柔ラカクナシ置ク急塲ノ切ハ是ニテ充分ナル

コアリ

總テ繩ノ流儀ニ依テ此位多者無シト故ドモ其法

ハ類ノ多イ丈ニテ長繩ヲ用ユルニアラズ故ニ此

度著ス處ハ實地必用之處ヲ著ス者ナレバ各先生

等ニテモ是ニテ充分ナルコヲ余ニ聞セラレタル故大略ヲ記載ハ此餘ノ

處ハ前ニ記シタル貳書ニアリ

Haya Nawa Zukai
Illustrated Explanation of Fast Tie : Illustration from Behind

This technique ties the criminal's thumbs on top of each other. After ensuring they are securely tied wrap the rest of the cord around his collar. This is done with a 7 Sun (21 centimeter) piece of string. This is shown in the illustration.

As was shown in the previous technique, when you do not have sufficient rope for a large group of criminals, strands of 5 Sun (15 centimeter) string can be used. Persons likely to encounter such situations should carry some thick pieces of silk cord of various lengths, 15, 21 and 30 centimeters long, as a matter of course. Soft pieces of cord are easy to store and can be immediately put to use if the situation calls for them.

These techniques using short pieces of string to restrain criminals are not common in Hojo Ryugi, Schools of Rope Bondage. That being said, there are many different methods. However, they are not meant to be used with longer arresting ropes. Since you may encounter a situation where you are arresting a large number of people, the author felt it necessary to include these examples. After discussion with the Sensei of various schools of martial arts, these very simplified instructions have been included. For further information I recommend reading the two books mentioned at the beginning of this chapter.

手錠縄圖解

盗賊及ヒ重罪ノ者ハ手錠ヲ掛タル上ニ縄ヲ以テ縛シ此ノ腰ニ巻附ルモアリ第一圖ノ如ク左手ニ縄ヲ一ツ環ニシテ内ヨリ一ニト二ニ二ツ巻

Illustration 1

第一圖

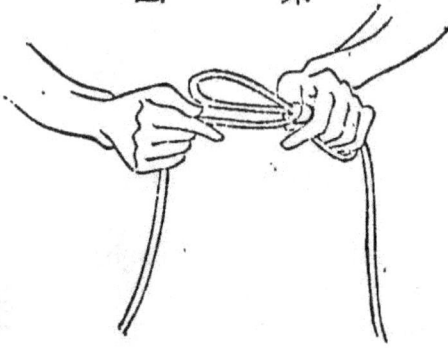

Illustration 2

第二圖

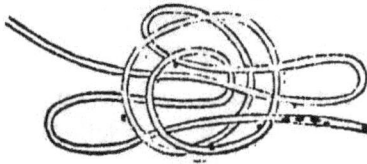

Illustration 3

第三圖

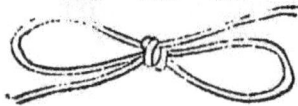

第二圖ノ如クニナシ両手首ニ掛ケテ縄ノ両端ヲ持テ引ハ第三圖ノ如ク二締ル故手錠ノ下ニテ垣根結ニナスベシ都合ニ依テ二重三重巻コトモアリト知ベシ

Tejo Nawa Zukai : **Illustrated Guide to Tying Rope Handcuffs**

Rope handcuffs should be tied on thieves and criminals that have committed serious crimes in addition to regular binding. As illustration 1 shows, hold the rope with your left hand and make a loop. Inside that loop make first one then another loop so the tie looks as shown in illustration 2. Place one of the criminal's hands in each loop and pull the ends of the rope tight. The final rope handcuffs are shown in illustration 3. The two ends of rope should be tied in a fence knot below the hands. Depending on the situation, a double or triple loops can be used to secure the hands.

活法圖解注意

總テ活法ヲ施スニハ武術ヲ好ム者ハ專要ノ處ナリ最モ此術ハ寳地ニ當
リ其時ニ此術ノ役ニ相不成ハ其功無シ故ニ活法學ニハ常々ノ稽古ニア
リ毎日氣合ヲ入勇氣ヲ以テ充分ノ働ヲナスコ肝要ナリ只働トハ死者ニ
向ヒ胸動氣ヲナシ水死、絞縊、落馬等各速死ヲ助ケルニハ我ガ心ヲ靜メ
其死者ノ救助法ニ注意スベシ最モ柔術家ノ秘密口傳是レアリ活法ノ内
ニモ其死ニ依テ術ノ施シ方モアリ是最モ注意肝要ノ第一トナス柔術先
生ニ於テモ目錄以上免許ニ相成秘傳ノ殺活ヲ許ス者成ハ此書ヲ見タ斗
リニテ死者ニ向ヒ充分ノ稽古ヲ成サル内ニナスベカラズ著者堅ク禁シ
置者也余ハ幾度モ死シ又他人ヲ活法ニテ蘇生サセシコ數回アリ最
モ初心ノ者ニハ其術ノ功ナキハ氣合ノ不入ト手足ノワザ通シ難キガ爲
ナリ故ニ柔術稽古ヲ專務シテ活法術ヲ施ス稽古ヲナスベシ
次回ニハ死相ノ圖解ヨリ各流ノ活法圖解有之ト雖余ガ前述ニアル殺活
自在接骨療法柔術生理書ニハ當身活法術ヲ詳細ニアリ尚此書ニモ有リ

Kappo Zukai Chu-i :
Cautions Regarding the Illustrated Guide to Resuscitation

Overview

Up until now the only people who trained in Kappo, Resuscitation, were those with long years of training in Jujutsu. The reason for this is each victim requires a different approach. After judging the situation the practitioner will decide whether resuscitation will be possible or not and which technique should be used. Thus Kappo Gaku, the Study of Resuscitation, was incorporated into the regular teaching done in Jujutsu schools. The techniques are only effective if you train diligently, by focusing your martial spirit and challenging yourself on a daily basis.

Procedure

The first step in resuscitation is checking for a pulse in the chest. Your patients will be victims of drowning, hanging, falls from horses and other sudden impacts so you must remain calm. It is important to focus on determining which rescue technique to apply. Traditionally these techniques were all closely guarded secrets that were never written down and only transmitted orally from the master of a Jujutsu school to a student. The practitioner would carefully consider the victim's condition before deciding which technique to apply and how to apply it. This evaluation is the most important step in resuscitation.

History

Typically the head of a Jujutsu school would only teach resuscitation techniques to a student that had been ranked higher than Mokuroku (see the following page for information on this term,) so clearly they were kept very secret.

The author strictly forbids the application of these techniques on a victim until the reader has thoroughly read and understood the contents and practiced extensively. I have used these techniques on several occasions to revive the dead or other victims, however there have been cases where novices have failed to revive victims. The reason for this was a combination of lack of determination when performing the technique along with a hesitant pressure with the hands and feet. Thus, in my opinion, the reader should commit themselves to Jujutsu training and within that setting study the arts of resuscitation.

Other Resources

The next illustrations will cover the topic of determining whether a patient is alive or dead in addition to introducing resuscitation techniques from each school of martial arts. I would like to mention these techniques as well as Atemi, striking points, are all exhaustively detailed in Sakkatsu Jizai Sekkotsu Chiho Jujutsu Seiri Sho, *Free Application of Life and Death : Essentials of Jujutsu Therapy. (Complete Book of Jujutsu Therapy* published in 1896.)

Note :

The rank of Mokuroku

A Mokuroku is a list or catalog that indicates a practitioner has achieved a certain level of proficiently. The reason it is a "list" or "catalogue" is because it only contains the names of a section of techniques. The Mokuroku document itself is fairly simple, it typically has the name of the school, the word Mokuroku at the beginning, the list of techniques in the center and ends with official signatures and stamps. There may be a list of all the heads of the school in chronological order until the current head.

This is part of a Kage School Mokuroku from the 15th ~16th century. The small dashes at the top indicate each technique. Under is the name of the technique and a single notation.

This is a chronological list of the heads of the school, starting on the right, with the diety Marishiten. A line traces the passing of the school from one to the next to the current head.

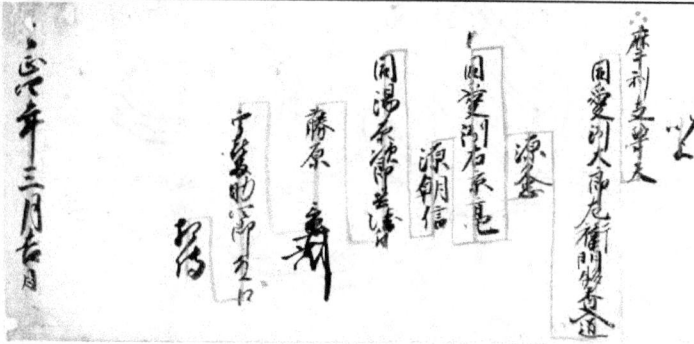

死相瞻圖解

總テ即死首絞リ高キ處ヨリ落馬等ニテ氣絶
ナシタル者假死者ニ向ヒタレハ直ニ右脇
ノ處ヘ圖ノ如ク左膝ヲ突キ右膝ヲ立（足ノ利クカ何レニテ）死者ノ両手ヲ臍ノ下ノ所ニ重テ靜ニ死
者ヲ視ルベシ是最モ肝要ノ處ナリ第一死者
ノ樣子ヲ能ク閲其上ニテ打處ヲ視テ眼中ヲ
視ル口ヲ開キ水月熱ミアルカ脇ノ下ニ脉ク
有バ必蘇生ナス者ナリ其廻リヲ靜ニナシ左
リ手ヲ枕ニナシテ抱起シテ次活ヲ施スベシ
最モ首絞死ハ糞ヲ垂レ居ルカ舌先ヲ嚙バ術
ノ功ナキ故ニ先ニ死相ヲ視ルコテ勉ムベシ

Shisotan Zukai
The Essential Points for Determining Life and Death

Those that have suddenly dropped dead, hung themselves, fallen from a high place, fallen off a horse or other such incident have lapsed into a state of unconsciousness called Kari Shi-sha, "mostly dead."

First approach the victim from his right side. As the illustration shows plant your left knee on the ground while keeping your right knee upright. If your opposite leg is your dominant one, then you can have your right knee on the ground and your left leg upright. Take the "mostly dead" victim's hands and place them on top of each other below his navel. Next, take a long quiet look at your patient. Careful observation of the "deceased" is the most important point of the technique. Look at the injury or point of impact, as well as looking into the eyes and opening the mouth. Check for heat around Suigetsu, the solar plexus, and for a pulse under the armpit. If either are present then the person can be brought back to life. Wrap your left arm around the back of his head like a pillow and pull him upright to perform resuscitation.

If a person who has hung themselves has fouled their pants or bitten off their tongue then the resuscitation will not have any effect. Thus it is essential to determine if the victim is truly alive or truly dead.

Note :
The 1898 version of this book ends here. The 1899 version includes several more resuscitation techniques. It's not clear if they forgot to include them in the earlier version or decided to add them later.

活 法 圖 解　睡氣活　左正面寫圖

此活ハ死相ヨリ引續キ死者ヲ抱起シテ
右膝ヲ突キ左膝ヲ立圖ノ如ク死者ノ右
側ニ構ヘ左掌ニテ死者ノ誘活ノ處ニ押
當右手ハ指ヲ中指ニ食指ヲ重テ揷畫ノ
如ニナシ臍ノ下ニ二寸餘ノ處毛ノ生基ノ
處ニ押當死者ノ顏ヲ下ヨリ視上ゲテ左
掌ト右指先ニ力チ入（エイ）ト右手ヲ押
込樣ト左ハ活處ヲ押テ死者ノ顏ヲ仰向
ク樣ニナシテ術ヲ施スベシ最
モ此活法ハ元各流ヨリ出タル
處ナレ共久富先生尙修正ノ處
モアリ他ハ前書ニ仝シテ故ニ
之ヲ畧ス

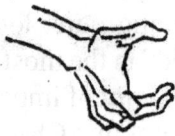

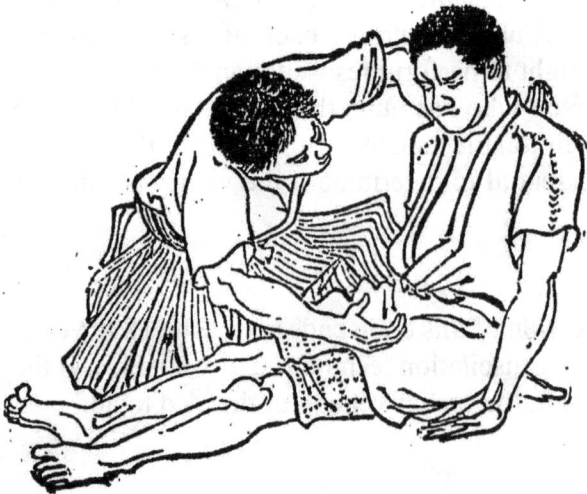

***Kappo Zukai* : Illustrated Guide to Resuscitation**
***Tanyu Katsu* : Bladder Meridian Resuscitation**
The Victim is in front of you to your left

Tanyu Katsu 胆兪活 Bladder Meridian Resuscitation is done after determining if the victim is alive or dead. Plant your right knee on the ground and keep your left leg upright as you raise the victim as shown in the illustration. You are positioned on the right side of the victim so use your left hand to push on Yuh-Katsu, the resuscitation point on the back.

Cross your middle finger over the back of your index finger as shown in the detail illustration. Push on the point approximately 2 Sun (6 centimeters) below the navel. This is approximately where pubic hair begins to grow. While looking up at the victim's face press into this point. Put power in the palm of your left hand and into the fingers of your right hand. With an exhalation of *Ei!* push in with your left palm into the resuscitation point on the victims back and, at the same time, push with the fingers of your right hand. You should do all this while looking up into the victim's face.

About this resuscitation technique :
The same technique can be found in many schools of Jujutsu, however this version was adapted by Hisatomi Sensei. There are no significant differences in the other schools so no further comments are necessary.

活法圖解　心臓活法　横左寫圖

此活モ種々施シタル後チ尚術ヲ施スベシ最モ
此活ハ大事ノ活故口傳アリト云ヘドモ實地ニ
能クキクナルガ充分之練麿ノ上ニテ施スベシ
死者ニ股ガリ右膝ヲ突キ左ヲ立膝トナシ兩手
ノ指ヲ組合シ死者ノ首ヲ抱ヘ込兩肘ヲ死者ノ
水落ノ處ニ押當（エイ）ト聲ヲ死者ニ我カ意氣
ヲ死者ニ呼吸チウッス様ニナスベシ此活ハ成
丈後ニナスベシ最モ練麿ノ上ニハ極意ノ處故
早ク蘇生ナス者ナリ天神眞揚流ナゾニテハ目
錄以上ニテ師ヨリ許ス者ナレバ其心得ニテ實
地ニ施スベシ

活法圖解　裏活左正面寫圖

此泹ハ死者ヲ腹歩ニ寐シテ股ノ處ニ跨ガリ左
膝ヲ突キ右膝ヲ立死者ノ裏腰ノ處ヘ挿畫如ク
兩掌ヲ以テ點ノ處ニテ大腿骨ノ上ノ處ヲ押當

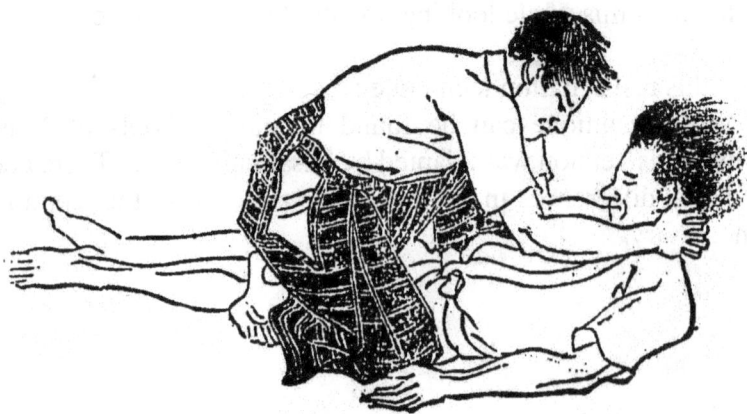

Kappo Zukai : **Illustrated Guide to Resuscitation**
Shinzo Kappo : **Heart Resuscitation**
Illustration from the side, you are on the left

Overview

This resuscitation method should only be used after various other methods have been tried. Traditionally, this resuscitation technique is kept secret and only transmitted orally from the master to the student, however since it has proven to be a highly effective method for reviving people, I will introduce it here. That being said, you need to practice it extensively before applying it in a real situation.

Procedure

First you want to lay the victim on his back and straddle him with your right knee on the ground. Next, reach behind his neck and interlace your fingers so you are holding the victim's neck up. Place your elbows on Mizu-ochi, the solar plexus. With an exhalation of *Ei!* push into the victim's solar plexus with both elbows. The feeling should be as if you are pushing with your spirit to restart the victim's respiration.

Caution

You should do this resuscitation after attempting other forms first. Since this represents the ultimate peak of resuscitation the recuperative effects can be quite immediate. Typically in schools like Tenshin Shinyo School, the master will not instruct a student in these arts until they have reached Mokuroku proficiency. You should keep this in mind when training or applying this art.

共力ノ結ビ最前押ノ最解書図ニヨリ是處注意ス最總是處秘於秘ニチナル依ハ
氣合腹ロ口共ケ活膝活一骨ニ活着者ナ法ノ語細訓ナ傳モスヘ故ニ讀ナ略ハニ
氣下入ニト下此ルノ同心脳ニ細ラ意ナチドリヨアヘ活ハ其大ヘ観レ記生ラシ
上聲モトノ大解詳總注ノ意ナ二ラテ

Kappo Zukai : **Illustrated Guide to Resuscitation**
Ura Katsu : **Resuscitation from the Back**
Illustration from the side, you are on the left

This technique is done after placing the victim face-down. As you step towards the thighs of the victim, lift your Hakama up so it is out of the way. Plant your left knee on the ground and keep your right knee up. Place the palms of both hands on the small of the victim's back, just above the hip-bone. This is shown in the detailed illustration.

Focus your power in the middle of your palms, as shown in the detail illustration, and push into the area just above the hip bone. When pushing you should gather your mental energy, put power in your lower abdomen, close your mouth firmly and push upward with a shout of *Ei!* Typically this resuscitation method is done in conjunction with the previous Heart Resuscitation.

Further Reading

Additional information and illustrations of the proper pressure point can be found in my *Complete Book of Jujutsu Therapy*.

The techniques presented in this book contain all the proper instructions and cautions however, this book only presents heavily abbreviated version. Thus these are not the full teachings of the resuscitation arts. For further information and details that have been kept secret and only transmitted orally up to now, please consult the *Complete Book of Jujutsu Therapy.*

Note : Inoguchi uses the word Daitai Kotsu, or thighbone/ upper-leg bone to describe what is clearly the point directly above the hip-bone. I have translated the section to reflect this, however, note that the author seems to consider the top of the hip-bone to be the top of the leg-bone.

Kenpo : An Illustrated Instructor's Manual

End

www.ingramcontent.com/pod-product-compliance
Lightning Source LLC
Chambersburg PA
CBHW061212220326
41599CB00025B/4613

9 781950 959181